# Contents

# Letter From the Principal Deputy Assistant Attorney General of the Office of Justice Programs

In February 2015, the President sent a memorandum to federal agencies directing those that have a role in the deployment of unmanned aircraft systems — or UAS — to ensure that effective safeguards are in place to protect citizen privacy, civil rights, and civil liberties. The memorandum underscores the potential benefits of UAS technology, cost savings and safety chief among them. At the same time, promoting the responsible use of this technology is critical, such that agencies that use UAS have sound policies in place governing the collection, use, and retention of data; that information is not used in a discriminatory manner; and that agencies are transparent in how they are using these systems.

In order for law enforcement agencies to do their jobs most effectively and most fairly, it is vital that they do their work in close association with communities, and in a way that garners the trust and support of the citizens they serve. Our mission at the Office of Justice Programs is to support state and local public safety efforts through funding, training, and research and development. One of our top priorities is to help law enforcement and other criminal and juvenile justice agencies build strong, respectful, and lasting relationships with the people they are sworn to protect and serve, and to embed a mindset of trust and respect into every facet of that work.

Having the confidence of the community is a critical public safety matter, and technology has the potential to facilitate trust, not to mention improve safety, both for citizens and for law enforcement officers. While UAS technology has the potential to be a useful public safety tool, the decision of whether and how it should be adopted is one that every law enforcement agency must make for itself — and that decision should be made in coordination with community members. These community discussions should consider privacy, civil rights, and civil liberties concerns.

The goal of the convening in August 2015 and this report is not to endorse the use of unmanned aircraft systems, or even to encourage any particular piece of equipment, but to describe how best to ensure law enforcement is building on innovative UAS efforts while keeping necessary safeguards in place. The convening's goal was to produce a blueprint for how law enforcement agencies can use unmanned aircraft systems most effectively, fairly, and transparently. The experience and insights of the experts who participated in the convening have been crucial to making sure NIJ is providing state and local law enforcement the best, most up-to-date guidance. The discussions were forward-thinking and advanced our efforts to create this report for implementing a UAS program that addresses all the stakeholders' interests.

Beth McGarry
Principal Deputy Assistant Attorney General
Office of Justice Programs

# Letter From the Director of the National Institute of Justice

The mission of the National Institute of Justice is to advance justice by applying scientific methods and strategies to solve real-world problems. Our job is to combine science with the experiences of criminal justice and public safety professionals to provide the best advice to help the nation deal with diverse contemporary topics and challenges related to criminal justice.

This report covers a number of emerging issues and concerns on the use of unmanned aircraft systems by state, local, and tribal law enforcement and public safety agencies discussed at a two-day convening of experts hosted by NIJ in Washington, D.C. on August 12 and 13, 2015. This meeting was in response to President Obama's Memorandum, *Promoting Economic Competitiveness While Safeguarding Privacy, Civil Rights, and Civil Liberties in Domestic Use of Unmanned Aircraft Systems*, published earlier in 2015. NIJ greatly appreciates the participation of each attendee at the meeting so that we can help each other understand how best to proceed in dealing with issues about unmanned aircraft systems.

The first day included a plenary session where a number of presentations by the Department of Justice, the Department of Homeland Security, local law enforcement agencies, civil liberties organizations, aviation associations, and legal experts offered perspectives on unmanned aircraft use in operations, concerns about surveillance, and other topics. Throughout the two days, the discussion touched on several dimensions pertinent to the successful use of UAS technology, such as community engagement; procurement and funding considerations; training needs; implementation requirements; retention issues; policy concerns; and prosecutor, civil rights, and privacy advocate interests.

Following the plenary session, two concurrent expert panels met on the first day. The first panel was composed of law enforcement and public safety leaders, experienced practitioners, and national policy leaders who contributed ideas and content related to policy for public safety agencies. The second panel was composed of community, privacy, and civil liberties advocates who contributed ideas and content to broaden input and ensure transparency. On the second day, the two panels reconvened and jointly discussed recommendations for the implementation and use of UAS technologies.

It is always encouraging to watch diverse stakeholders come together to share their perspectives on thorny issues. We hope this report provides useful information for law enforcement and public safety agencies considering UAS as a part of their operations. We also hope this document serves as an accessible introduction to the issues for the interested public who may have questions, concerns, or curiosity about the impact this technology may have on their communities.

Nancy Rodriguez, Ph.D.
Director
National Institute of Justice

# Introduction

In February 2015, President Obama issued a memorandum to federal agencies directing that agencies with a role in the deployment of unmanned aircraft systems (UAS) follow policies to ensure that effective strategies are in place to protect citizens' privacy, civil rights, and civil liberties.[1] The Presidential Memorandum also addresses the potential benefits of UAS technology, including cost savings and reduction of risks to human life. The U.S. Department of Justice (DOJ) wants to promote the responsible use of UAS technology by encouraging state and local law enforcement agencies that use this technology to implement sound policies governing data collection, use, and retention that is both transparent and non-discriminatory.

To further this goal, the Office of Justice Programs/National Institute of Justice (OJP/NIJ) convened a meeting in Washington, D.C. in August 2015, bringing together a comprehensive group of public safety stakeholders and aviation experts to discuss how best to ensure that they are building on innovative UAS technology efforts while keeping necessary safeguards in place. Representatives of community advocacy groups also participated in the meeting and their input was essential to this process. In order for law enforcement agencies to do their jobs most effectively and most fairly, it is vital that they do their work in close association with communities in a way that garners the trust and support of the citizens those agencies serve.

Having the confidence of communities is a critical public safety matter. Communities are safer when trust is greater, and people are more likely to comply with the law and cooperate with authorities if they believe the process is fair and free of bias. Public safety and community confidence go hand-in-hand. OJP has a national initiative to build community trust and justice and is conducting demonstrations and evaluations of innovative methods that advance procedural justice and evidence-based community engagement strategies. OJP's goal is to embed a mindset of trust and respect in every aspect of the criminal justice system.

The goal of the 2015 convening was not to endorse the use of UAS technologies or to encourage any particular type of equipment. While UAS technology has the potential to be a useful public safety tool, each public safety agency has to decide for itself whether this technology will have a positive impact on its mission. The decision should be made in

---

[1] *Presidential Memorandum: Promoting Economic Competitiveness While Safeguarding Privacy, Civil Rights, and Civil Liberties in Domestic Use of Unmanned Aircraft Systems,* accessible at https://www.whitehouse.gov/the-press-office/2015/02/15/presidential-memorandum-promoting-economic-competitiveness-while-safegua and included in Appendix A of this document.

concert with the community that each agency serves. Discussions should involve community stakeholders, and the privacy, civil rights, and civil liberties of citizens should be given major consideration in the development and implementation of any UAS program.

The desired outgrowth of the meeting is the development of a basic blueprint on how law enforcement agencies can use UAS most effectively, fairly and transparently. The experience and insights of the meeting participants was crucial to the goal of giving the field the best and most up-to-date guidance. This report reflects the comments and issues raised by the UAS convening participants and includes some essential recommendations for successfully implementing a UAS program.

# Limitations

When reading this report, it is important to keep in mind the limitations of the methods that were used to elicit content. This report relies largely on the opinions of a diverse group of subject matter experts. Although the best possible effort was made to identify a representative set of experts, by no means were all potential sources of expertise exhausted. The participants in this convening provided an honest and insightful view of both the concerns and potential solutions associated with implementation of a public safety UAS program within a community. While the technologies and related issues seem to change on a daily basis, the hope is that this report will serve as a useful primer for public safety agencies considering implementing UAS technologies.

# Chapter 1. Benefits and Risks of UAS Technologies

As UAS technology has made rapid advances, so has interest within the government as to their potential uses in furtherance of public safety missions.

According to the Bureau of Justice Statistics, only about 350 law enforcement agencies in the U.S. had aviation programs in active use.[2] This number has remained small due to the substantial cost of operating fixed and rotary wing aircraft. The Department of Justice's (DOJ) Office of the Inspector General (OIG) September 2013 report noted that small UAS weighing up to 55 pounds can have much lower operational and maintenance costs than the manned aircraft typically used by law enforcement.[3] The report cites an estimate that UAS can operate at $25 per hour versus $650 per hour for manned aircraft operations. As the technology matures, this minimal cost may shrink even further, making UAS available to even more law enforcement agencies as a lower cost alternative to manned aircraft.

In addition to the cost factor, UAS systems are capable of performing a number of missions that are simply too dangerous or are beyond the technical capabilities of manned aircraft. Current UAS public safety uses include response to and assessment of hazardous materials (HAZMAT) spills and incidents, explosive ordnance disposal (EOD) incidents, search and rescue missions, crime scenes, barricaded subject surveillance, active-shooter incidents, execution of search warrants, disaster response and recovery, training support, and assistance requests from local and state fire authorities. Other potential law enforcement uses of these platforms are:

- Forensic photography/crime scene mapping.

- Damage assessment.

- Public safety communications enhancement.

---

[2] Lynn Langdon, *Aviation Units in Large Law Enforcement Agencies, 2007,* Bureau of Justice Statistics, Washington, DC, July 2009, http://www.bjs.gov/content/pub/pdf/aullea07.pdf.

[3] Office of the Inspector General, Audit Division, *Interim Report on the Department of Justice's Use and Support of Unmanned Aircraft Systems,* Report 13-37, U.S. Department of Justice, Washington, DC, September 2013, https://oig.justice.gov/reports/2013/a1337.pdf.

- Emergency response/disaster management/post-disaster assessment.

- Operational planning.

- Border patrol.

- Aerial photography and security patrolling of critical infrastructure.

- 3-D mapping of major transportation accidents.

- Aerial surveillance.

- Public safety/life preservation missions.

- Alarm response (roof checks, inaccessible fenced-in areas).

- Crowd monitoring (under carefully defined circumstances so as not to infringe on the First Amendment rights of those gathered).

- For situational awareness during critical incidents such as active shooters.

UAS technology has the potential to minimize the amount of force that would otherwise be required in a situation. For example, one agency used UAS to find a suspect in a cornfield within minutes, rather than use a canine search. Officers warned the suspect that if he didn't surrender they would deploy the canine. He surrendered quickly with no injury to himself or officers.

The fire services are also very interested in the potential shown by these systems, which can give a fire commander real-time situational awareness on large fires, both wildland and structural. The addition of a thermal or hyperspectral sensor to the UAS can assist a fire agency with determining "hot spots" and "flare-ups" at a fire scene.

While the potential for these systems to extend the capabilities of both law enforcement and fire services is great, the task of implementing them in a way that is understood and accepted by the communities where they are to be used is considerable. Part of the problem is that small UAS platforms are often referred to as "drones," which people associate with military, weaponized technology. There is also a heightened sensitivity to privacy concerns connected with aerial surveillance. Communities have expressed concerns about privacy with the implementation of public safety UAS. A March 2014 study polled members of the public regarding their perceptions of using a ground-based camera instead of a UAS-mounted camera for certain functions.[4] Generally, the study found that the surveyed population would accept using ground-based cameras for certain activities, such as enforcing traffic laws, but would exhibit much higher levels of concern about a UAS-mounted camera gathering the same information.

[4] Kerry G. Herron, Hank C. Jenkins Smith, and Carol L. Silva, *US Public Perspectives on Privacy, Security, and Unmanned Aircraft Systems, Center for Risk and Crisis Management,* Center for Applied Social Research, University of Oklahoma, Norman, Oklahoma, March 2014, http://crcm.ou.edu/pvcy2014/report.pdf.

Some of the potential use cases for unmanned aircraft the general public identified as raising significant privacy and civil liberties concerns include using UAS to:

- Monitor traffic for issuing traffic citations.

- Pursue suspects beyond "visual line of sight" (VLOS).

- Conduct surveillance over large groups of people.

- Drop objects, such as tear gas.

- Exercise force via weaponized UAS.

- Serve as weapon target (laser) designators.

As additional law enforcement agencies acquire UAS, it is clear that early and honest dialogue with the community is an absolute necessity if broad public support and acceptance is to be achieved. Appendix A includes a number of guidance and resource documents to assist agencies in establishing a safe, legal, and acceptable environment for UAS usage.

# Chapter 2. Privacy and Civil Liberties Issues

Although not explicitly mentioned in the text of the U.S. Constitution, the U.S. Supreme Court has found that several amendments have created various privacy rights, and these rights form the basis for various other legal and policy protections. Additionally, privacy rights have been created through statute, including through The Privacy Act of 1974, 5 U.S.C. § 552a. In the context of using UAS, the Fourth Amendment to the U.S. Constitution provides certain guarantees of privacy with respect to protecting certain areas from government search or intrusion.[5] The U.S. Supreme Court has held that, for an area to be entitled to protection, a mere individual, subjective expectation of privacy is not enough — there must be an expectation of privacy that "society is prepared to recognize as reasonable."[6]

Discussions about how far the right to privacy extends in the unmanned aircraft context often center on how best to protect the rights of individuals against unwarranted intrusions by law enforcement agencies seeking to collect, maintain, use, and disclose imagery obtained through use of a UAS. In these discussions, it is important to note that the U.S. Supreme Court, since 1967, has held that areas and activities exposed to the public are not protected under the Fourth Amendment, as it is not objectively reasonable to expect your actions to be private when done in view of the public.[7] With respect to aerial surveillance, the U.S. Supreme Court has repeatedly ruled that law enforcement is not required to obtain a search warrant, as individuals do not have a Fourth Amendment right to be free from warrantless aerial surveillance.[8]

---

[5] *Skinner v. Railway Labor Executives' Assn.,* 489 U.S. 602, 613-614 (1989) (the Fourth Amendment "'guarantees the privacy, dignity, and security of persons against certain arbitrary and invasive acts by officers of the Government,' without regard to whether the government actor is investigating crime or performing another function.").

[6] *See, e.g., Katz v. United States,* 389 U.S. 347, 361 (1967); *Oliver v. United States,* 466 U.S. 170, 177-178 (1984).

[7] *See, e.g., Katz, supra,* at 351 ("What a person knowingly exposes to the public, even in his own home or office, is not a subject of [Fourth] Amendment protection."); *California v. Ciralo,* 476 U.S. 207, 213 (1986) ("The Fourth Amendment . . . has never been extended to require law enforcement officers to shield their eyes when passing by a home on public thoroughfares. Nor does the mere fact that an individual has taken measures to restrict some views of his activities preclude an officer's observations from a public vantage point where he has a right to be and which renders the activities clearly visible.").

[8] *See, California v. Ciralo,* 476 U.S. 207, 215 (1986) ("The Fourth Amendment simply does not require the police traveling in the public airways . . . to obtain a warrant in order to observe what is visible to the naked eye."); *Dow Chemical Co. v. United States,* 476 U.S. 227, 239 (1986) (holding that the taking of aerial photographs of private property by government officials from navigable airspace is not a search under the Fourth Amendment); *Florida v. Riley,* 488 U.S. 445 (1989).

Powerful technologies have always had the potential to shift the balance of power between individuals and the government, including public safety or law enforcement agencies. Most law enforcement agencies that are currently using UAS technology are doing so for specific, narrowly circumscribed purposes, such as operations at crime scenes or emergency responses to active shooter situations. But the major concern of privacy community advocacy groups is the possible use of UAS for routine mass surveillance, or "persistent surveillance," by which the government could record the comings and goings of vehicles and pedestrians over a wide area in cities and towns across the country. Accordingly, participants in this convening discussed a number of related topics, including the impact that government use of UAS may have on civil rights and civil liberties; comparison of using unmanned aircraft to collect and retain personally identifiable information (PII) to other methods, such as manned aircraft, body-worn cameras, dash cameras and pole cameras; and the public safety aspects of UAS use compared with more conventional surveillance or evidence-gathering methods.

All participants in the convening agreed that there is a need to be proactive about crafting rules, regulations, and sub-regulatory guidance regarding governmental use of UAS. Well-informed policymakers should strike a careful balance between the interests of law enforcement and the promotion of individual privacy, civil rights, and civil liberties, rather than waiting for litigation in the courts to flesh out the details of permissible practices in narrow, case-by-case legal determinations.

There is also some agreement that regulation of surveillance by aircraft should be the same for manned and unmanned aircraft, with the understanding that there may be some additional regulatory oversight required for unique unmanned aircraft issues. As with all other investigative technologies, law enforcement and public safety agencies should, on a regular basis, develop, review, and update policies and regulations to govern the use of new technologies, law enforcement approaches, and potential privacy concerns. Finally, public safety officials should work collaboratively with their communities in order to find a middle ground on these issues and in a manner that builds trust between the parties. One way to do this is for agencies to establish public advisory bodies to help develop a concept of operations for UAS that obtains community consensus and support while also permitting law enforcement and public safety agencies to leverage this emerging technology.

Although some aspects were not unique to UAS, questions arose concerning data collection and retention by the government:

(1) When, and for what purpose, should law enforcement be able to collect data/PII?

(2) Should collected data be retained/stored and, if so, for how long?

(3) Is persistent mass surveillance ever acceptable (either by law enforcement or non-law enforcement agencies)?

(4) Is there is a difference between live viewing and recording (ability to analyze recorded images)? Is it more acceptable to view data via UAS without recording it?

(5) Should collected and retained data ever be disseminated?

(6) Should guidelines for data collection, retention, use, and dissemination be platform-agnostic (i.e., the same for UAS as for manned aircraft, pole cameras, body-worn cameras, etc.)?

One specific concern regards governmental use of UAS for non-law-enforcement purposes, such as to advance fire service activities, fish and wildlife service operations, search and rescue missions, or emergency evacuations. Some fear that collecting and retaining data for such seemingly benign purposes may also infringe civil liberties.

From the perspective of the American Civil Liberties Union (ACLU), for example, when it comes to institutions, incentives matter. Law enforcement has a difficult job, which is sometimes perceived to be made easier by collecting as much information as technically possible — an incentive that might prove inconsistent with adequate privacy protections. Indeed, because UAS can be more cost-effective than manned aircraft, the potential ubiquity of cheap, unmanned surveillance devices may lead to more significant potential for abuse of individuals' personal information.

Some members of the law enforcement community feel that certain state legislatures have overreacted to such concerns, enacting measures that are overly restrictive and do not fully consider all of the implications of the legislation.

On the other hand, privacy community advocacy groups believe that there is reason to be suspicious of government use of UAS technology, whether at the local, state, or federal level. Law enforcement and privacy community advocacy groups disagree on whether use of UAS is appropriate during protest gatherings, and privacy community advocates fear persistent surveillance by government and wanton collection and storage of PII.

But there is also consensus among public safety and privacy community advocates that warrantless use of UAS technology should be used to accomplish important missions in ways that do not encroach on people's privacy. For example, UAS can offer a unique perspective from the sky that can change the way in which first responders and other government personnel manage emergency situations, such as a hazardous materials spill, hurricane evacuation, crime scene, automobile accident or search for a missing person.

In promoting such worthy missions, participants focused on accountability, which prompted questions regarding the specific circumstances and authorities under which the government should be allowed to use UAS. For instance, how long should data be retained? Are the privacy interests different if a UAS operator is only viewing a scene via live video, rather than recording the scene for later review and analysis? Many of these questions remain unanswered and do not appear to be within the scope of existing rulemaking by federal agencies. For example, the Federal Aviation Administration (FAA) issues regulations to ensure the safety of civil aviation within the National Airspace System (NAS) and grants waivers to permit government agencies to conduct certain public aircraft operations in ways that do not jeopardize the safety of other aircraft. The FAA's rules typically are concerned with flight safety rather than the collection of data by airborne sensor packages.

To summarize, on one side, privacy advocates are afraid of rogue and excessive use of UAS technology and want stronger controls. On the other side, representatives of the law enforcement community oppose the imposition of unnecessary controls that will inhibit their ability to conduct their public safety mission and to quickly respond to exigent, life-threatening situations. All parties concur that the technology has value; the challenge for all agencies seeking to use UAS will be formulating policy that accommodates the needs of law enforcement and public safety while also ensuring appropriate privacy and civil liberty protections based on existing law and regulation.

Additionally, law enforcement officials should consider how they are going to use UAS technologies while not diminishing trusted relationships with the communities that they serve. For example, if UAS is used in a missing person search, should police be able to use the images or video recorded during the mission to prosecute a serious crime that they happen to discover during the search?

# Chapter 3. Considerations for the Implementation of UAS Technologies

As law enforcement agencies contemplate implementing UAS capabilities to enhance their delivery of public safety services, the following three considerations should be addressed:

**I. Actions Internal to the Agency**

a. All members of the agency should be briefed on the technical capabilities and intended operational use of the UAS technology. This creates understanding of a common message regarding UAS when interacting with the public. For example, "The UAS technology will only be used two ways, for training and at a defined incident perimeter, such as a fire or crime scene. There will be no 'free' flights without a specific and defined mission."

b. It is essential that the agency identify the actual need for UAS technologies to support its public safety mission. It is important to remember the mantra, "The mission should drive the technology; the technology should not drive the mission." Accordingly, an agency should:

- Identify the specific missions for which the UAS technology will be used. If the agency cannot identify missions that are unique to UAS, the agency should reconsider initiating a UAS program. (See the "Technology Decision Tool" in Appendix A.)

- Determine whether the specific UAS technology that best matches the agency's intended use is commercially available. In making this determination, agencies can consult with the U.S. Department of Homeland Security (DHS), NIJ's Aviation Technology Program, the Airborne Law Enforcement Association (ALEA), the Association for Unmanned Vehicle Systems International (AUVSI), and other agencies that are using UAS. The agency should also determine if there is another method for performing the mission, perhaps at a lower cost.

- Analyze whether applicable federal, state and local laws, regulations, and policies allow the agency's intended use of UAS. Such analysis should include research regarding rules governing airspace or other geographical restrictions.

- Assess or develop the parameters for sharing UAS technology with other agencies, conducting inter-agency operations, or otherwise supporting other agencies via mutual aid requests. For example, the FAA has promulgated rules and regulations concerning the shared and support role of public aircraft operations.

- Evaluate whether the agency can support the effort by reference to technology, funding, storage, maintenance, and training considerations. What type of staff should conduct the UAS operation? Who will be responsible for UAS in the agency? If the agency has an aviation unit, will UAS operations fall within the scope of that unit's responsibilities?

c. In developing privacy policies, it is essential for an agency's leadership to fully understand the complex legal environment in which UAS operate. Agency officials should:

- Understand how privacy and Fourth Amendment issues apply to the agency's intended use of UAS. Model privacy policies and operational procedures are available to use as templates for the agency's own policy and procedures, but must be individualized to the agency's needs and the requirements of the jurisdiction. If the agency is federal or has received federal funding for its UAS program, check the February 2015 Presidential Memorandum to ensure that that agency's policies are compliant.

- Bear in mind that certain states and localities have imposed more stringent requirements on law enforcement's use of UAS technology than those imposed by federal law.

- Develop written and transparent policies to help alleviate community concerns about UAS technology, to the extent that law enforcement and public safety considerations permit. It is important that the agency's legal advisors be integrated into the policy-development process, as well as the oversight and auditing of UAS operations.

## II. Community Engagement

a. It is important for the agency to engage the community early in the process of implementing UAS technologies. The following steps to effective community engagement may prove helpful:

- Consistent with the needs of law enforcement and public safety, invite the community to be involved in the exploratory process of selecting and implementing UAS technology, which helps to combat the perception of government secrecy and promote transparency.

- Engage early. Community engagement is most effective at the beginning of the process of implementing a UAS program. With some members of the public suspicious of the government using new surveillance technologies, the more transparent the process, the higher the likelihood of a successful launch. For example, if the agency is considering employing UAS such as small quad-rotor helicopters ("quadcopters"), explaining the size, capabilities, and limitations of that particular unmanned aircraft may help citizens better understand the goals of the program and promote trust.

- Create a permanent community advisory panel on the implementation of new technologies, such as UAS, drawing from a wide cross-section of the population. The size of the advisory group should be proportional to the size and population of an agency's jurisdiction and should only include citizens of the jurisdiction. Jurisdictions should be careful not to appoint non-citizens to be members and be careful about permitting individuals who represent advocacy groups to be a part of this group. While inviting representatives from advocacy groups may help inform the panel, they should not be voting members to ensure that the views of the actual citizens are reflected. A properly assembled citizen advisory panel can gauge resistance and help allay community fears.

- Establish an annual audit of the UAS program and, consistent with the needs of law enforcement and public safety, publicly disseminate the report.

b. Getting the honest message out requires taking the following actions:

- Become knowledgeable about UAS technology and how it will be used. Implementation of the technology should correspond to a clearly defined mission, which should be articulated clearly to the community before a purchase decision.

- Ensure the agency has the economic and technical aspects correct about the UAS technology. Agency officials will be expected to be familiar with facts such as costs, maintenance requirements, and intended use.

- Involve the agency's Public Information Officer (PIO) in the engagement process. Once a decision is made to acquire and deploy UAS technology, hold media events to educate the public.

- Invite media to training exercises so that the press has a better understanding of UAS technology and how it will be used.

- Be prepared to listen to members of the community and their concerns, and make adjustments as necessary.

## III. Implementation

a. The agency should consider taking these steps with respect to actual implementation of a UAS program:

- Begin a working relationship with the FAA for public aircraft operations. A guide for working with the FAA, understanding applicable FAA rules, and obtaining a Certificate of Waiver or Authorization (COA) for public aircraft operations is available at http://www.faa.gov/about/office_org/headquarters_offices/ato/service_units/systemops/aaim/organizations/uas/coa/.

- Identify the agency's funding source for both the initial purchase and continued operation of UAS technology.

- Perform due diligence before purchasing UAS technologies to ensure that vendors can support the technical claims they make regarding their unmanned aircraft and its sensors. Does a particular UAS really fly for 30 minutes? What does its warranty cover? (See "Things to Consider for Unmanned Aircraft Systems (UAS) Procurement" in Appendix A).

- Determine if the agency's aircraft, if necessary or required, has received the necessary certifications from the FAA.

- Conduct recurring UAS training once the agency implements the UAS technology. This training should not only include actual flight operations and simulated missions, but also legal and policy training on UAS and aerial surveillance, including a discussion of privacy and constitutional issues, and updates on any new laws, regulations, or other sources of applicable law that might affect the agency's use of UAS.

- Develop an assessment program to ensure that the UAS technology is meeting the promised capabilities.

b. The following steps are relevant to the agency's management of the UAS operational team and UAS program:

- Carefully select the staff of a UAS program based on their aviation knowledge and training (or willingness to get the necessary training). Staff of the UAS program should have a record of positive interaction with the community and the media. They also should have a demonstrated record in adhering to the agency's rules and policies so that they will follow agency rules for the operation of UAS technologies. This will enhance the safety of UAS operations and decrease the agency's risk of legal liability.

- Understand that there are limitations to using UAS technologies, including restrictions in aviation law and other laws and regulations regarding the operation of manned and unmanned aircraft. For example, management should understand why a UAS operator may not be permitted to operate a UAS under poor weather conditions or near airports.

- Determine whether the UAS technology is a cost-effective tool for the agency. Logs of flights, issues, costs, and benefits should be kept so that an annual study of the efficacy of the program can be conducted. The results should be disseminated publicly in order to promote the transparency of the program.

- Ensure that the agency's insurance covers UAS and UAS operations. If not, the agency should investigate the availability of commercial insurance policies.

- Be prepared to defend decisions that the agency makes regarding the purchase, implementation, and operational use of UAS technology, both within the government and externally to the community that the agency serves. Having good policies and practices in place will make it easier for an agency to defend its actions.

- Share and learn from best practices in UAS operations.

# Chapter 4. Lessons Learned

Following is a summary of lessons learned, in checklist form, for implementing a UAS program:

❏ Clearly identify the need for the technology and the specific missions that the agency wants to accomplish with a UAS, and why a UAS is the right choice for the missions. Remember the mantra, "The mission should drive the technology; the technology should not drive the mission."

❏ Fully understand the complex legal environment in which these machines will operate. Determine whether federal, state and local laws, regulations, and policies allow use of UAS in the ways envisioned. Involve the agency's legal advisors at the beginning of the process.

❏ Determine whether the agency can support the UAS effort, including evaluating the technology, funding, storage, maintenance, and training.

❏ Carefully work through the authorization to fly process with the FAA. Begin with the FAA web page (http://www.faa.gov/about/office_org/headquarters_offices/ato/service_units/systemops/aaim/organizations/uas/coa/) and consult with other agencies that have completed this process.

❏ Involve the department's Public Information Officer early in discussions about a UAS program.

❏ Directly and effectively involve the community/constituents at the beginning of the process. When dealing with a public that is inherently suspicious of UAS technologies, the more transparent the process, the higher the likelihood of a successful launch of a UAS program.

❏ Consider establishing a permanent community advisory panel on the implementation of UAS and other new technologies. An advisory panel can help agencies work collaboratively with the community to build consensus and support.

❏ Do not purchase UAS technology without being able to justify the need. Ensure the agency acquires the right UAS technology, if it is currently commercially available, to support the missions envisioned.

❏ Develop written policies defining acceptable missions, data collection, and data/ personally identifiable information (PII) storage/retention.

❏ Ensure that the agency's UAS program is as transparent as possible. Establish clear policies that the agency is able and prepared to defend. Be prepared to defend decisions regarding the purchase, implementation, and operational use of UAS technology.

❏ Develop an assessment program to ensure that the UAS technology is meeting the vendor's promised capabilities.

❏ Share and learn from best practices in UAS operations.

# Appendix A. Guidance and Resource Documents

# Presidential Memorandum

Federal Register / Vol. 80, No. 34 / Friday, February 20, 2015 / Presidential Documents
Page 9355

Memorandum of February 15, 2015

**Promoting Economic Competitiveness While Safeguarding Privacy, Civil Rights, and Civil Liberties in Domestic Use of Unmanned Aircraft Systems**

**Memorandum for the Heads of Executive Departments and Agencies**

Unmanned Aircraft Systems (UAS) technology continues to improve rapidly, and increasingly UAS are able to perform a variety of missions with greater operational flexibility and at a lower cost than comparable manned aircraft. A wide spectrum of domestic users—including industry, private citizens, and Federal, State, local, tribal, and territorial governments—are using or expect to use these systems, which may play a transformative role in fields as diverse as urban infrastructure management, farming, public safety, coastal security, military training, search and rescue, and disaster response.

The Congress recognized the potential wide-ranging benefits of UAS operations within the United States in the FAA Modernization and Reform Act of 2012 (Public Law 112–95), which requires a plan to safely integrate civil UAS into the National Airspace System (NAS) by September 30, 2015. As compared to manned aircraft, UAS may provide lower-cost operation and augment existing capabilities while reducing risks to human life. Estimates suggest the positive economic impact to U.S. industry of the integration of UAS into the NAS could be substantial and likely will grow for the foreseeable future.

As UAS are integrated into the NAS, the Federal Government will take steps to ensure that the integration takes into account not only our economic competitiveness and public safety, but also the privacy, civil rights, and civil liberties concerns these systems may raise.

By the authority vested in me as President by the Constitution and the laws of the United States of America, and in order to establish transparent principles that govern the Federal Government's use of UAS in the NAS, and to promote the responsible use of this technology in the private and commercial sectors, it is hereby ordered as follows:

**Section 1.** *UAS Policies and Procedures for Federal Government Use.* The Federal Government currently operates UAS in the United States for several purposes, including to manage Federal lands, monitor wildfires, conduct scientific research, monitor our borders, support law enforcement, and effectively train our military. As with information collected by the

Federal Government using any technology, where UAS is the platform for collection, information must be collected, used, retained, and disseminated consistent with the Constitution, Federal law, and other applicable regulations and policies. Agencies must, for example, comply with the Privacy Act of 1974 (5 U.S.C. 552a) (the "Privacy Act"), which, among other things, restricts the collection and dissemination of individuals' information that is maintained in systems of records, including personally identifiable information (PII), and permits individuals to seek access to and amendment of records.

(a) *Privacy Protections.* Particularly in light of the diverse potential uses of UAS in the NAS, expected advancements in UAS technologies, and the anticipated increase in UAS use in the future, the Federal Government shall take steps to ensure that privacy protections and policies relative to UAS continue to keep pace with these developments. Accordingly, agencies shall, prior to deployment of new UAS technology and at least every 3 years, examine their existing UAS policies and procedures relating to the collection, use, retention, and dissemination of information obtained by UAS, to ensure that privacy, civil rights, and civil liberties are protected. Agencies shall update their policies and procedures, or issue new policies and procedures, as necessary. In addition to requiring compliance with the Privacy Act in applicable circumstances, agencies that collect information through UAS in the NAS shall ensure that their policies and procedures with respect to such information incorporate the following requirements:

(i) *Collection and Use.* Agencies shall only collect information using UAS, or use UAS-collected information, to the extent that such collection or use is consistent with and relevant to an authorized purpose.

(ii) *Retention.* Information collected using UAS that may contain PII shall not be retained for more than 180 days unless retention of the information is determined to be necessary to an authorized mission of the retaining agency, is maintained in a system of records covered by the Privacy Act, or is required to be retained for a longer period by any other applicable law or regulation.

(iii) *Dissemination.* UAS-collected information that is not maintained in a system of records covered by the Privacy Act shall not be disseminated outside of the agency unless dissemination is required by law, or fulfills an authorized purpose and complies with agency requirements.

(b) *Civil Rights and Civil Liberties Protections.* To protect civil rights and civil liberties, agencies shall:

(i) ensure that policies are in place to prohibit the collection, use, retention, or dissemination of data in any manner that would violate the First Amendment or in any manner that would discriminate against persons based upon their ethnicity, race, gender, national origin, religion, sexual orientation, or gender identity, in violation of law;

(ii) ensure that UAS activities are performed in a manner consistent with the Constitution and applicable laws, Executive Orders, and other Presidential directives; and

(iii) ensure that adequate procedures are in place to receive, investigate, and address, as appropriate, privacy, civil rights, and civil liberties complaints.

(c) *Accountability.* To provide for effective oversight, agencies shall:

  (i) ensure that oversight procedures for agencies' UAS use, including audits or assessments, comply with existing agency policies and regulations;

  (ii) verify the existence of rules of conduct and training for Federal Government personnel and contractors who work on UAS programs, and procedures for reporting suspected cases of misuse or abuse of UAS technologies;

  (iii) establish policies and procedures, or confirm that policies and procedures are in place, that provide meaningful oversight of individuals who have access to sensitive information (including any PII) collected using UAS;

  (iv) ensure that any data-sharing agreements or policies, data use policies, and record management policies applicable to UAS conform to applicable laws, regulations, and policies;

  (v) establish policies and procedures, or confirm that policies and procedures are in place, to authorize the use of UAS in response to a request for UAS assistance in support of Federal, State, local, tribal, or territorial government operations; and

  (vi) require that State, local, tribal, and territorial government recipients of Federal grant funding for the purchase or use of UAS for their own operations have in place policies and procedures to safeguard individuals' privacy, civil rights, and civil liberties prior to expending such funds.

(d) *Transparency.* To promote transparency about their UAS activities within the NAS, agencies that use UAS shall, while not revealing information that could reasonably be expected to compromise law enforcement or national security:

  (i) provide notice to the public regarding where the agency's UAS are authorized to operate in the NAS;

  (ii) keep the public informed about the agency's UAS program as well as changes that would significantly affect privacy, civil rights, or civil liberties; and

  (iii) make available to the public, on an annual basis, a general summary of the agency's UAS operations during the previous fiscal year, to include a brief description of types or categories of missions flown, and the number of times the agency provided assistance to other agencies, or to State, local, tribal, or territorial governments.

(e) *Reports.* Within 180 days of the date of this memorandum, agencies shall provide the President with a status report on the implementation of this section. Within 1 year of the date of this memorandum, agencies shall publish information on how to access their publicly available policies and procedures implementing this section.

**Sec. 2.** *Multi-stakeholder Engagement Process.* In addition to the Federal uses of UAS described in section 1 of this memorandum, the combination of greater operational flexibility, lower capital requirements, and lower operating costs could allow UAS to be a transformative technology in the commercial and private sectors for fields as diverse as urban infrastructure management, farming, and disaster response. Although these opportunities will enhance American economic competitiveness, our Nation must be mindful of the potential implications for privacy, civil rights, and civil liberties. The Federal Government is committed to promoting the responsible use of this technology in a way that does not diminish rights and freedoms.

(a) There is hereby established a multi-stakeholder engagement process to develop and communicate best practices for privacy, accountability, and transparency issues regarding commercial and private UAS use in the NAS. The process will include stakeholders from the private sector.

(b) Within 90 days of the date of this memorandum, the Department of Commerce, through the National Telecommunications and Information Administration, and in consultation with other interested agencies, will initiate this multi-stakeholder engagement process to develop a framework regarding privacy, accountability, and transparency for commercial and private UAS use. For this process, commercial and private use includes the use of UAS for commercial purposes as civil aircraft, even if the use would qualify a UAS as a public aircraft under 49 U.S.C. 40102(a)(41) and 40125. The process shall not focus on law enforcement or other noncommercial governmental use.

**Sec. 3.** *Definitions.* As used in this memorandum:

(a) "Agencies" means executive departments and agencies of the Federal Government that conduct UAS operations in the NAS.

(b) "Federal Government use" means operations in which agencies operate UAS in the NAS. Federal Government use includes agency UAS operations on behalf of another agency or on behalf of a State, local, tribal, or territorial government, or when a nongovernmental entity operates UAS on behalf of an agency.

(c) "National Airspace System" means the common network of U.S. airspace; air navigation facilities, equipment, and services; airports or landing areas; aeronautical charts, information, and services; related rules, regulations, and procedures; technical information; and manpower and material. Included in this definition are system components shared jointly by the Departments of Defense, Transportation, and Homeland Security.

(d) "Unmanned Aircraft System" means an unmanned aircraft (an aircraft that is operated without direct human intervention from within or on the aircraft) and associated elements (including communication links and components that control the unmanned aircraft) that are required for the pilot or system operator in command to operate safely and efficiently in the NAS.

(e) "Personally identifiable information" refers to information that can be used to distinguish or trace an individual's identity, either alone or when combined with other personal or identifying information that is linked or linkable to a specific individual, as set forth in Office of Management and Budget Memorandum M–07–16 (May 22, 2007) and Office of Management and Budget Memorandum M–10–23 (June 25, 2010).

**Sec. 4.** *General Provisions.*

(a) This memorandum complements and is not intended to supersede existing laws and policies for UAS operations in the NAS, including the National Strategy for Aviation Security and its supporting plans, the FAA Modernization and Reform Act of 2012, the Federal Aviation Administration's (FAA's) Integration of Civil UAS in the NAS Roadmap, and the FAA's UAS Comprehensive Plan.

(b) This memorandum shall be implemented consistent with applicable law, and subject to the availability of appropriations.

(c) Nothing in this memorandum shall be construed to impair or otherwise affect:

   (i) the authority granted by law to an executive department, agency, or the head thereof; or

   (ii) the functions of the Director of the Office of Management and Budget relating to budgetary, administrative, or legislative proposals.

(d) Independent agencies are strongly encouraged to comply with this memorandum.

(e) This memorandum is not intended to, and does not, create any right or benefit, substantive or procedural, enforceable at law or in equity by any party against the United States, its departments, agencies, or entities, its officers, employees, or agents, or any other person.

(f) The Secretary of Commerce is hereby authorized and directed to publish this memorandum in the Federal Register.

THE WHITE HOUSE,
*Washington, February 15, 2015*
[FR Doc. 2015–03727
Filed 2–19–15; 2:00 pm]
Billing code 3295–F5

# Department of Justice Policy Guidance[1]
# Domestic Use of Unmanned Aircraft Systems (UAS)

## *INTRODUCTION*

The law enforcement agencies of the Department of Justice ("the Department") work diligently to protect the American people from national security threats, enforce our nation's laws, and ensure public safety. In doing so, these agencies use a wide variety of investigative methods. Some of these methods have been in use for decades; others are relatively new and rely on technological innovation. In all cases, investigations and other activities must be conducted consistent with the Constitution and the laws of the United States—and with our commitment to protecting privacy and civil liberties.

In recent years, Unmanned Aircraft Systems (UAS)[2] have emerged as a viable law enforcement tool. UAS have been used to support kidnapping investigations, search and rescue operations, drug interdictions, and fugitive investigations. While they are, in many ways, similar to the manned aircraft that have been in use for many years, they have the potential to provide law enforcement with additional flexibility and yield life-saving benefits. UAS also have the potential to be cost-effective in a time of shrinking government resources. For these reasons, UAS are likely to come into greater use.

As technology advances and enhances our ability to use these new tools, it is important to continue to assess how we use them. A Departmental working group[3] has studied the Department's use of UAS over the last several years and has considered how the technology is likely to evolve in the near future. This policy guidance flows from the working group's

---

[1] This policy guidance is intended only to improve the internal management of the Department of Justice. It is not intended to and does not create any right, benefit, trust, or responsibility, whether substantive or procedural, enforceable at law or equity by a party against the United States, its departments, agencies, instrumentalities, entities, officers, employees, or agents, or any person, nor does it create any right of review in an administrative, judicial or any other proceeding.

[2] "Unmanned Aircraft System" means an unmanned aircraft (an aircraft that is operated without the possibility of direct human intervention from within or on the aircraft) and associated elements (including communication links and components that control the unmanned aircraft) that are required for the pilot or system operator in command to operate safely and efficiently in the National Airspace System. For purposes of this policy, reference to "UAS" includes all onboard sensor equipment.

[3] The Department's working group was led by the Office of Legal Policy and included the Department's Chief Privacy and Civil Liberties Officer and representatives of the Bureau of Alcohol, Tobacco, Firearms and Explosives, the Criminal Division, the Office of Community Oriented Policing, the Civil Rights Division, the Office of the Deputy Attorney General, the Drug Enforcement Administration, the Federal Bureau of Investigation, the National Security Division, the Executive Office for United States Attorneys, the Office of Justice Programs, the Office of Privacy and Civil Liberties, the United States Marshals Service, and the Office of the Chief Information Officer.

discussions and sets forth principles that will apply Department-wide. This policy also applies to all instances in which Department components use UAS to support Federal agencies and/or State and Local law enforcement agencies.

This guidance will help ensure that the Department continues to carry out its law enforcement and national security missions while respecting individuals' privacy, civil rights, and civil liberties. It will also help ensure an appropriate level of accountability and transparency. This policy guidance does not replace, and is complementary to, the Federal Aviation Administration rules and regulations that control each and every UAS deployment and help ensure the safe operation of all aircraft, including UAS. This policy guidance is also consistent with the Presidential Memorandum, "Promoting Economic Competitiveness While Safeguarding Privacy, Civil Rights, and Civil Liberties in Domestic Use of Unmanned Aircraft Systems," issued by President Barack Obama on February 15, 2015.

## RESPECT FOR CIVIL RIGHTS AND CIVIL LIBERTIES

Respect for civil rights and civil liberties is a core tenet of our democracy. In executing the Department's law enforcement and national security missions, personnel must rigorously support and defend the Constitution and continue to uphold the laws, regulations and policies that govern our activities and operations.

As with all investigative methods, UAS must be operated consistent with the U.S. Constitution. The Fourth Amendment protects individuals from unreasonable searches and seizures and generally requires law enforcement to seek a warrant in circumstances in which a person has a reasonable expectation of privacy. Moreover, Department personnel may never use UAS solely for the purpose of monitoring activities protected by the First Amendment or the lawful exercise of other rights secured by the Constitution and laws of the United States. Department personnel may never use UAS to engage in discrimination that runs counter to the Department's policies on race, ethnicity, gender, national origin, religion, sexual orientation, or gender identity. Department personnel must also be trained to understand and abide by all relevant federal legal standards applicable to the use of UAS, and to seek advice from legal counsel as necessary.

In addition, UAS may only be used in connection with properly authorized investigations and activities. Statutory authorities, the Attorney General's Guidelines, and other relevant agency policies and guidance define the scope of authorized investigations and activities and require regular supervisory review and approval. UAS must continue to be used within the context of these existing safeguards.

Further, even within the context of properly authorized activities, personnel often must choose among different investigative methods that are operationally sound, reasonable, and effective, but may be more or less intrusive relative to individuals' privacy and civil liberties. Prior to using UAS, Department personnel must assess the relative intrusiveness of the proposed use of UAS, and balance it against the particular investigative need.[4] This is both

---

[4] In assessing the intrusiveness of UAS and the investigative need, personnel must consider factors such as whether the subject enjoys a reasonable expectation of privacy relative to the proposed use of UAS, the scope of the information sought, the scope of the proposed use of UAS, the risk of disclosure to the subject, the seriousness of the crime or national security threat, the strength and significance of the information to be obtained, the efficiency of the method and alternative means available, the amount of information already known about the subject, and the operational security needs of the investigation.

a logical process and an exercise in judgment, but the overall principle remains: in deciding whether to use UAS, Department personnel must consider and, if reasonable based on the facts and circumstances of the investigation, use the least intrusive means to accomplish an operational need.

## PROTECTION OF PRIVACY

The Department operates under a set of rules, policies, and laws that control the collection, retention, dissemination, and disposition of records that contain personally identifiable information. For example, the Privacy Act contains provisions on unauthorized use and disclosure of information about individuals, and imposes civil penalties on agencies and criminal penalties on agency personnel for violations of applicable requirements. As with personally identifiable information collected in the course of any investigation, these authorities apply to information collected via UAS. Consistent with applicable existing laws and requirements, the Department's use of UAS shall include the practices identified below.

As noted above, the Department shall only collect, use and disseminate information obtained from UAS for an authorized purpose. The Department shall not retain information collected using UAS that may contain personally identifiable information for more than 180 days unless retention of the information is determined to be necessary for an authorized purpose or is maintained in a system of records covered by the Privacy Act.

Data collected by UAS that is retained must be safeguarded in accordance with applicable Federal laws, Executive Orders, directives, policies, regulations, standards, and guidance. These authorities ensure that Department personnel with access to such data follow practices that are consistent with the protection of privacy and civil liberties. Use of all Department information systems may be monitored, recorded, and subject to audit, and unauthorized collection, retention, or dissemination of data is prohibited. Further, the Department has procedures in place to review, investigate, and address privacy and civil liberties complaints.

Senior Component Officials for Privacy in agencies using UAS must conduct annual privacy reviews of their agency's use of UAS to ensure compliance with existing laws, regulations, and Department policy, and to identify potential privacy risks. They must also, where appropriate, make recommendations to ensure that UAS will continue to be used in a manner consistent with the U.S. Constitution and all applicable laws, regulations, and policies, including those protecting privacy and civil liberties.

## ACCOUNTABILITY

The Department promotes accountability by requiring its personnel to accept responsibility for the actions they undertake—and to evaluate the potential consequences of their decisions. The Department imposes codes of conduct to guide employees in the use of all investigative methods, including UAS. As with the use of any technology, there must continue to be mechanisms to hold the Department and its employees accountable.

Part of accountability is ensuring that Department personnel are appropriately trained and supervised. Department personnel whose responsibility it is to manage, supervise, maintain, fly, and/or otherwise use UAS must receive training on this policy and the underlying policies incorporated herein.

Moreover, approval authority for the use of UAS will be set at an appropriate and consistent level across the Department. At a minimum, each time UAS are deployed, approval must be granted (1) at the Assistant Special Agent in Charge-or-equivalent level at the relevant field office, and (2) by an executive level supervisor within the agency's aviation support unit or a designated executive level supervisor at the agency's headquarters. Additionally, since the Department may only operationally deploy UAS in connection with authorized investigations or activities, supervisors must ensure that the underlying investigations themselves have been authorized consistent with applicable guidelines and other Department policies.

Finally, federal records must be captured, managed, and retained in a manner consistent with the Federal Records Act and all other applicable authorities. As with federal records collected by other investigative tools, components are obligated to retain UAS-collected data in accordance with applicable records retention schedules.

## ONGOING POLICY MANAGEMENT

As UAS technology evolves and improves, it is important that the Department continue to have adequate information about its use to ensure strategic alignment and proper evaluation of the Department's policy. To that end, this policy imposes certain new requirements.

Each component that uses UAS must designate a point of contact through which field offices will report the information outlined below to the component's headquarters and Department leadership on the use of UAS on an ongoing basis.

In addition, Department agencies that use UAS must report annually to the Deputy Attorney General on the use of UAS. The report should incorporate privacy reviews, as well as the number of UAS operational deployments (not including training or research and development flights) conducted• during the reporting period and a brief description of types or categories of missions flown along with the number of each type of mission. Additionally, to the extent the agency sought assistance from, or provided assistance to, another federal, state, local, or tribal agency during the relevant time period, the number of these operational deployments and a brief description of types or categories of missions flown along with the number of each type of mission should also be provided.

Components that have not previously disclosed any UAS operations as part of these annual reporting requirements, or that have discontinued UAS use for the duration of an annual reporting period, must notify the Deputy Attorney General prior to initiating or re-introducing UAS operations.

Department leadership will continue to engage in meaningful review of UAS as the technology advances. To facilitate this review, a standing committee comprised of a broad range of Department components will meet twice a year to evaluate any policy or regulatory changes that may be needed as a result of innovations or developments in UAS technology.

## TRANSPARENCY

Rigorous adherence to the requirements set forth in this policy is not enough—to be successful in our law enforcement and national security missions, we must continue to facilitate relationships of trust with the communities we serve. Enhancing our transparency about agency operations, including how we operate UAS, creates an informed citizenry and greater confidence in the Department's decision-making.

Education of the public can enhance the Department's ability to fulfill its missions and serve the American people. As appropriate, while not revealing information that might compromise law enforcement or national security needs, the Department will update its website to reflect its current policy on UAS on an ongoing basis, and will provide a general summary of UAS operations conducted by the Department during the previous year, including a brief description of types or categories of missions flown and the number of times the Department provided assistance to other federal, state, local and tribal agencies or entities.

**Federal Aviation Administration Certificates of Waiver or Authorization**

**Federal Aviation
Administration**

# Certificates of Waiver or Authorization
# (COA)

COA s an author zat on ssued by the A r Traffic Organ zat on to a pub c operator for a spec fic UA act v ty After a comp ete app cat on s subm tted FAA conducts a comprehens ve operat ona and techn ca rev ew If necessary prov s ons or m tat ons may be mposed as part of the approva to ensure the UA can operate safe y w th other a rspace users In most cases FAA w prov de a forma response w th n 60 days from the t me a comp eted app cat on s subm tted

To better support the needs of our customers FAA dep oyed a web-based app cat on system The UAS COA On ne System (https //ioeaaa faa gov/oeaaa/Welcome jsp) prov des app cants w th an e ectron c method of request ng a COA App cants w need to obta n an account n order to access the on ne system

P ease ema the FAA/UAS office at 9-AJR-36-UAS@faa gov w th any quest ons or for more nformat on regard ng Cert ficates of Wa ver or Author zat on

**Quick Links:**

- Letter to COA Ho ders Statutory Requ rement to Reg ster UAS (November 5 2014) (www faa gov/uas/resources/uas regu at ons po cy/med a/Reg strat on etter pdf) (PDF)
- UAS COA On ne System (http //ioeaaa faa gov/)
- Pub c y Re eased COAs (www faa gov/uas/resources/foia responses/ )
- Frequent y Asked Quest ons (www faa gov/uas/faqs/)
- COA: Samp e App cat on (PDF 1 5 MB) (https://www.faa.gov/about/office org/headquarters offices/ato/service units/systemops/aaim/ organizations/uas/media/COA%20Sample%20Application%20v%201-1.pdf)

Page last modified August 19 2016 8 21 58 AM EDT

Th s page was or g na y pub shed at:
http://www faa gov/about/office org/headquarters offices/ato/serv ce un ts/systemops/aa m/organ zat ons/uas/coa/

**International Association of Chiefs of Police Recommended Guidelines for the Use of Unmanned Aircraft**

INTERNATIONAL ASSOCIATION OF CHIEFS OF POLICE

AVIATION COMMITTEE

*Recommended Guidelines for the Use of Unmanned Aircraft*

**BACKGROUND:**

Rapid advances in technology have led to the development and increased use of unmanned aircraft. That technology is now making its way into the hands of law enforcement officers nationwide.

We also live in a culture that is extremely sensitive to the idea of preventing unnecessary government intrusion into any facet of our lives. Personal rights are cherished and legally protected by the Constitution. Despite their proven effectiveness, concerns about privacy threaten to overshadow the benefits this technology promises to bring to public safety. From enhanced officer safety by exposing unseen dangers, to finding those most vulnerable who may have wandered away from their caregivers, the potential benefits are irrefutable. However, privacy concerns are an issue that must be dealt with effectively if a law enforcement agency expects the public to support the use of UA by their police.

The Aviation Committee has been involved in the development of unmanned aircraft policy and regulations for several years. The Committee recommends the following guidelines for use by any law enforcement agency contemplating the use of unmanned aircraft.

**DEFINITIONS:**

1.  Model Aircraft — A remote controlled aircraft used by hobbyists, which is manufactured and operated for the purposes of sport, recreation and/or competition.

2. Unmanned Aircraft (UA) — An aircraft that is intended to navigate in the air without an on-board pilot. Also called Remote Piloted Aircraft and "drones."

3. UA Flight Crewmember — A pilot, visual observer, payload operator or other person assigned duties for a UA for the purpose of flight.

4. Unmanned Aircraft Pilot — A person exercising control over an unmanned aircraft during flight.

**COMMUNITY ENGAGEMENT:**

1. Law enforcement agencies desiring to use UA should first determine how they will use this technology, including the costs and benefits to be gained.

2. The agency should then engage their community early in the planning process, including their governing body and civil liberties advocates.

3. The agency should assure the community that it values the protections provided citizens by the U.S. Constitution. Further, that the agency will operate the aircraft in full compliance with the mandates of the Constitution, federal, state and local law governing search and seizure.

4. The community should be provided an opportunity to review and comment on agency procedures as they are being drafted. Where appropriate, recommendations should be considered for adoption in the policy.

5. As with the community, the news media should be brought into the process early in its development.

**SYSTEM REQUIREMENTS:**

1. The UA should have the ability to capture flight time by individual flight and cumulative over a period of time. The ability to reset the flight time counter should be restricted to a supervisor or administrator.

2. The aircraft itself should be painted in a high visibility paint scheme. This will facilitate line of sight control by the aircraft pilot and allow persons on the ground to monitor the location of the aircraft. This recommendation recognizes that in some cases where officer safety is a concern, such as high risk warrant service, high visibility may not be optimal. However, most situations of this type are conducted covertly and at night. Further, given the ability to observe a large area from an aerial vantage point, it may not be necessary to fly the aircraft directly over the target location.

3. Equipping the aircraft with weapons of any type is strongly discouraged. Given the current state of the technology, the ability to effectively deploy weapons from a small UA is doubtful. Further, public acceptance of airborne use of force is likewise doubtful and could result in unnecessary community resistance to the program.

4. The use of model aircraft, modified with cameras, or other sensors, is discouraged due to concerns over reliability and safety.

## OPERATIONAL PROCEDURES:

1. UA operations require a Certificate of Authorization (COA) from the Federal Aviation Administration (FAA). A law enforcement agency contemplating the use of UA should contact the FAA early in the planning process to determine the requirements for obtaining a COA.

2. UA will only be operated by personnel, both pilots and crew members, who have been trained and certified in the operation of the system. All agency personnel with UA responsibilities, including command officers, will be provided training in the policies and procedures governing their use.

3. All flights will be approved by a supervisor and must be for a legitimate public safety mission, training, or demonstration purposes.

4. All flights will be documented on a form designed for that purpose and all flight time shall be accounted for on the form. The reason for the flight and name of the supervisor approving will also be documented.

5. An authorized supervisor/administrator will audit flight documentation at regular intervals. The results of the audit will be documented. Any changes to the flight time counter will be documented.

6. Unauthorized use of a UA will result in strict accountability.

7. Except for those instances where officer safety could be jeopardized, the agency should consider using a "Reverse 911" telephone system to alert those living and working in the vicinity of aircraft operations (if such a system is available). If such a system is not available, the use of patrol car public address systems should be considered. This will not only provide a level of safety should the aircraft make an uncontrolled landing, but citizens may also be able to assist with the incident.

8. Where there are specific and articulable grounds to believe that the UA will collect evidence of criminal wrongdoing and if the UA will intrude upon reasonable expectations of privacy, the agency will secure a search warrant prior to conducting the flight.

## IMAGE RETENTION:

1. Unless required as evidence of a crime, as part of an on-going investigation, for training, or required by law, images captured by a UA should not be retained by the agency.

2. Unless exempt by law, retained images should be open for public inspection.

# U.S. Customs and Border Protection, Office of Air and Marine Unmanned Aircraft System Operations and Privacy Policy

OFFICE OF AIR AND MARINE
WASHINGTON, DC

OAM PROCEDURE NO. 2013-15
DATE: September 9, 2013

**Unmanned Aircraft System Operations and Privacy**

1.  PURPOSE. The purpose of this directive is to establish uniform policies, procedures, and guidelines for conducting U.S. Customs and Border Protection (CBP) Unmanned Aircraft Systems (UAS) surveillance operations while ensuring compliance with privacy law and policy.

2.  POLICY. It is Office of Air and Marine (OAM) policy that in accordance with U.S. law and consistent with this UAS surveillance operations directive, OAM agents and personnel may use CBP UAS to provide integrated and coordinated border interdiction and law enforcement support to homeland security missions; provide assistance, consistent with the prerogatives of the Department of Homeland Security (DHS) and CBP, to other federal, state, and local agencies in other law enforcement and emergency humanitarian efforts; provide airspace security for National Special Security Events; and combat smuggling and other cross-border violations.

    2.1 General Privacy Considerations.

        2.1.1 The video, still images, and radar images captured from a CBP UAS generally do not clearly identify individuals on the ground or elsewhere; however, these images may be associated with a particular individual, such as when an individual has been apprehended or is under custody.

        2.1.2 All biographical information obtained from apprehended individuals and any video or radar images of their movement collected from a CBP UAS that may reasonably be regarded as evidence in any kind of court or administrative proceeding must be retained in the appropriate law enforcement case management system, pursuant to the OAM Evidence Handling procedures and in compliance with the Privacy Act.

        2.1.3 Video, still images, and/or radar images collected from a CBP UAS as part of a natural disaster government response and/or emergency situation, which generally are not associated with any particular person, may be provided via real time feed to the Federal Emergency Management Agency and/or the corresponding state emergency operating center.

2.1.4 OAM personnel may use video, still images, and/or radar images collected from a CBP UAS to apprehend individuals and as evidence that may help prove a violation of law. Subject to applicable law, information collected from a CBP UAS may be shared with other federal, state, local, tribal, or foreign law enforcement agencies to assist them with the enforcement of the laws that they administer, and subject to compliance with confidentiality and other requirements (for example, those specified in 6 U.S.C. § 485 and 19 U.S.C. § 1628).

2.2 Data Minimization and Retention.

2.2.1 OAM personnel may use a CBP UAS to collect video, still images, and/or radar images only in support of an authorized mission and/or investigation.

2.2.2 OAM personnel may use a CBP UAS to collect, among other types of information, Personally Identifiable Information (PII) that is relevant and necessary in the course of an authorized mission and/or investigation.

2.2.3 Any retention of PII by OAM personnel must be performed in accordance with the requirements of any applicable Privacy Act System of Records Notice where the PII data is maintained.

2.2.4 In all cases, OAM personnel will minimize the long-term retention of any video, still images, and/or radar images collected from a CBP UAS. Storage media containing imagery collected from a CBP UAS that is not related to any particular case, operationally relevant, nor relate to a potential violation of law, will continue to be overwritten every 30 days or otherwise, in accordance with the OAM ground control station data storage system management procedures.

2.3 Data Use.

2.3.1 OAM personnel must use PII data only for the purposes for which such information was collected. OAM personnel may collect video, still images, and/or radar images, from a CBP UAS pursuant to their law enforcement authority, as part of their border security mission, or when flying a mission in support of another agency, provided that such other agency has the legal authority to request the use of a CBP UAS.

2.3.2 While the video resolution, radar mapping images collected from a CBP UAS are not sufficiently precise to permit actual identification of a person, such images or information could be associated with a particular individual when combined with the circumstances surrounding the activity revealed in the image or any other additional information obtained from such person.

2.3.3 Any sharing of information or data collected from a CBP UAS with other federal, state, local, tribal, or foreign law enforcement agencies to facilitate or assist them with the enforcement of the laws that they administer, must strictly comply with confidentiality and other requirements (for example, those specified in 6 U.S.C. § 485 and 19 U.S.C. § 1628), with OAM's Evidence Handling procedures, and the Privacy Act.

2.4 Data Quality and Integrity.

2.4.1 OAM personnel operating any CBP UAS are required to complete annual Privacy Awareness, Ethics, and CBP Code of Conduct training. Additionally, OAM UAS operators must successfully complete training on the proper operation of CBP UAS recording equipment, to guarantee and preserve the quality and integrity of any information that is collected from a CBP UAS, and which is intended to be used as evidence that may help prove a violation of law.

2.4.2 The Privacy Awareness training to be completed by OAM personnel operating any CBP UAS includes techniques to copy recorded evidence and a complete understanding of the OAM Evidence Handling policy. As with any information associated with a particular case file, once the images or videos are cross-referenced with, and included within records relating to, an ongoing investigation or case, they become covered by the system of records for that particular case file system and subject to the Privacy Act requirements of that system.

2.5 Data Security.

2.5.1 OAM personnel must protect video, still, and/or radar images captured from a CBP UAS, through appropriate security safeguards against risks to include loss, unauthorized access or use, destruction, modification, or unintended or inappropriate disclosure.

2.5.2 Live video and flight information sent from a CBP UAS is passed along an encrypted feed from the CBP UAS through the satellite relay to the ground control station. Any video or data that is transmitted, in real time via BigPipe to a closed system with restricted access is subject to access controls and an approval process requiring clearance by CBP system administrators to ensure that only authorized users with a need to know have access to the video feeds.

2.5.3 Real time video feeds transmitted from a CBP UAS are recorded by OAM personnel on a Digital Video Recorder (DVR). Any recorded images saved as evidence that may help prove a violation of law or for surveillance purposes must be handled in accordance with OAM Evidence Handling procedures, and in compliance with the Privacy Act. Any recorded images by OAM personnel that are intended to be used as evidence in a court or administrative proceeding must be kept in a locked container, segregated from other property and/or equipment.

2.5.4 Any video or images collected from a CBP UAS during an investigative operation containing sensitive surveillance or reconnaissance related data must be controlled and archived by Office of Intelligence and Investigative Liaison (OIIL) and may not be disclosed outside CBP without the express approval of the OIIL Collections Division Director.

2.5.5 Any information collected from a CBP UAS that is stored on the DVR that does not appear to constitute evidence that may help prove a violation of law or does not appear to be related to an ongoing investigation will be overwritten approximately every 30 days.

2.6 Operational Guidelines.

2.6.1 OAM personnel shall only conduct CBP UAS operations in support of authorized DHS/CBP border and homeland security missions, for training purposes, or when flying a mission in support of another federal, state, or local agency, provided that such other agency has the legal authority to request the use of a CBP UAS.

2.6.2 The OAM Assistant Commissioner shall be made aware of any requests for CBP UAS support made by any federal, state, or local agencies for non-routine operations.

2.6.3 OAM personnel shall comply will all applicable Federal Aviation Administration (FAA) rules and regulations while operating a CBP UAS.

2.6.4 OAM personnel are authorized to launch a CBP UAS in one of two ways:

a. Dynamically (during UAS flight in response to rapidly changing or officer safety event), or

b. Prior to launch, as a:

1. Joint Agency Request (e.g., Joint Field Command - Arizona, National Response Coordination Center, Joint Interagency Task Force - South, etc.), or

2. OIIL intelligence collection request based upon Homeland Security Standing Information Needs and CBP Priority Intelligence Requirements.

2.6.5 All requests for CBP UAS support must be coordinated through the Command Duty Officer. Prior to mission launch, requests for CBP UAS support must be coordinated in accordance with the current OAM Aviation Support Request Policy, through the Director, National UAS Operations, and as depicted below:

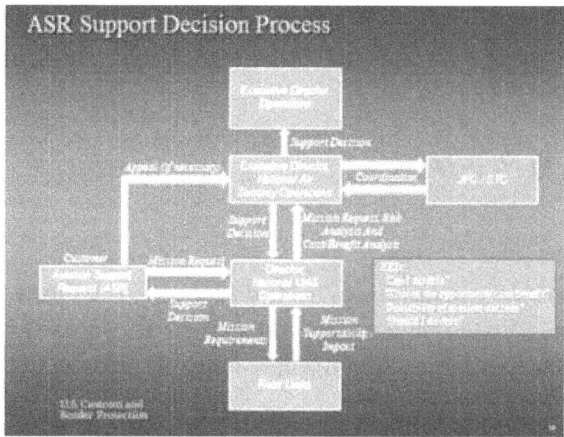

2.6.6 The deployment of a CBP UAS must be conducted on a priority basis; however, this commitment will not preclude the use of other CBP aviation resources in support of additional authorized DHS/CBP mission and/or investigation. The following mission sets are listed in order of priority:

- Tier 1: National DHS/CBP Missions

- Tier 2: CBP Missions

- Tier 3: Other Federal/State/Local Missions (Resources Permitting)

2.6.7 Specific missions listed in order of priority include:

1. DHS/CBP law enforcement officer needs assistance;

2. Any other persons need assistance in life-threatening situations;

3. Reported crimes in progress;

4. Investigative or other air support missions;

5. Routine mission support;

6. Maintenance test flights.

2.6.8 CBP UAS are not configured or authorized to carry munitions.

3. BACKGROUND. OAM protects the American people and the nation's critical infrastructure through the coordinated use of integrated air and marine assets to detect, interdict, and prevent acts of terrorism and the unlawful movement of people, illegal drugs, and other contraband toward or across the borders of the United States. To achieve its border surveillance and law enforcement missions, OAM employs several types of aircraft, including manned helicopters, fixed-wing aircraft, and UAS.

3.1 This policy supersedes any prior OAM policy to the extent that the prior policy is inconsistent with the content of this directive.

4. AUTHORITIES/REFERENCES.

4.1 Homeland Security Act, 6 U.S.C. § 111, et seq.

4.2 Department of Homeland Security Appropriations Act, 2013, Pub. L. No. 113-6, and successor appropriations thereto.

4.3 8 U.S.C. § 1357; 19 U.S.C. §§ 1581, 1590, 1644, 1644a, 1703.

5. RESPONSIBILITIES.

5.1 OAM supervisors will conduct CBP UAS operations in accordance with the OAM Aviation Operations Handbook and in compliance with applicable FAA regulations. OAM supervisors will also be responsible for documenting any investigative activity and incident reporting that occurs during the course of an aviation enforcement operation utilizing CBP databases.

5.2 The Air and Marine Operations Center (AMOC) will provide direct coordination between OAM agents and CBP assets to ensure proper investigative and operational deconfliction. In addition to the radar separation/advisory services provided by the primary Air Traffic Control facility for the region in which a CBP UAS is operating, AMOC will provide radar over-watch of all CBP UAS flights as a redundant safety precaution. AMOC will provide documentation utilizing the AMOC watch log to record any CBP aviation enforcement activity.

5.3 OAM agents should always seek local advice from the CBP Office of Chief Counsel (Associate/Assistant Chief Counsel) prior to sharing any video, still images, and/ or radar images or data collected from a CBP UAS with other federal, state, local, tribal, or foreign law enforcement agencies, and with respect to ongoing investigations or pending court cases, this consultation should also include the applicable U.S. Attorney's Office, through Associate/Assistant Chief Counsel, where appropriate.

6. NO PRIVATE RIGHT CREATED. This is an internal procedure of OAM and this procedure does not create or confer any rights, privileges, or benefits for any person or entity.

7. DISCLOSURE. This document contains law enforcement sensitive information that may be exempt from disclosure to the public pursuant to federal law. No part of this document shall be disclosed to the public without express authority from OAM.

Randolph D. Alles
Assistant Commissioner
Office of Air and Marine

# U.S. Department of Homeland Security Privacy Impact Assessment for the Aircraft Systems

Privacy Impact Assessment for the
Aircraft Systems

**DHS/CBP/PIA-018**
**September 9, 2013**

**Contact Point**
**Lothar Eckardt**
**Executive Director, National Air Security Operations**
**Office of Air & Marine**
**U.S. Customs and Border Protection**
**(202) 344-3950**

**Reviewing Official**
**Jonathan R. Cantor**
**Acting Chief Privacy Officer**
**Department of Homeland Security (202) 343-1717**

## Abstract

The Department of Homeland Security (DHS), U.S. Customs and Border Protection (CBP) employs several types of aircraft including manned helicopters and fixed-wing aircraft, and Unmanned Aircraft Systems (UAS) for border surveillance and law enforcement purposes. These aircraft are equipped with video, radar, and/or other sensor technologies to assist CBP in patrolling the border, conducting surveillance as part of a law enforcement investigation or tactical operation, or gathering raw data that may assist in disaster relief or responses to other emergencies. Video, images, and sensor data collected through these Aircraft Systems alone cannot be used to identify a person, but they may later be associated with a person as part of a law enforcement investigation or encounter with CBP officers or agents. DHS/CBP is conducting this Privacy Impact Assessment to evaluate the privacy impact of these technologies on persons.

# Introduction

The Department of Homeland Security (DHS), U.S. Customs and Border Protection (CBP) is responsible for guarding nearly 7,000 miles of land border the United States shares with Canada and Mexico and 2,000 miles of coastal waters surrounding the Florida peninsula and off the coast of Southern California. The agency also protects 95,000 miles of maritime border in partnership with the United States Coast Guard. To achieve these missions, CBP employs several types of aircraft, including manned helicopters and fixed-wing aircraft, and Unmanned Aircraft Systems (UAS) for border surveillance and law enforcement purposes. These aircraft are equipped with video, radar, and/or other sensor technologies to assist CBP in patrolling the border, conducting surveillance as part of a law enforcement investigation or tactical operation, or gathering raw data that may assist in disaster relief or other emergencies. This Privacy Impact Assessment (PIA) is necessary because the aircraft are equipped with technology that captures information that may be associated with persons whom CBP encounters.

## *Overview*

CBP employs several types of aircraft to achieve its mission objectives. All aircraft, manned or unmanned, have some type of imaging capability such as video, still images collection, and/or radar. The UAS differ from CBP's manned aircraft only in that the pilot controls the aircraft from the ground and the aircraft are capable of flying farther distances and longer hours continuously. All aircraft are owned and operated by the Office of Air and Marine (OAM); the Office of Intelligence and Investigative Liaison (OIIL) is responsible for processing, exploitation, and dissemination (PED) of imagery transmitted from aircraft.

CBP aircraft, both manned and unmanned, are used in the following scenarios: (1) to patrol the border; (2) to conduct surveillance for investigative operations; (3) to conduct damage assessment in disaster situations; and (4) in response to officer safety scenarios. While CBP also allocates its air assets in a manner that reflects this prioritization, CBP reviews and considers all requests for assistance. Lastly, CBP does not equip its aircraft with weapons. While the crew in all manned aircraft and the officers and agents onboard the aircraft during tactical missions do carry weapons, the various aircraft are not equipped with armaments.

### *Helicopters*

CBP operates several types of manned rotary-wing aircraft (helicopters) in support of its mission, notably, the American Eurocopter AS-350, Augusta Westland AW-139, Bell Huey UH1, and Sikorsky UH-60. CBP uses helicopters for observation, for tracking suspects and supporting ground units, aerial reconnaissance of moving objects and persons, external lift capability for seizures and equipment delivery, and tactical support and transportation for law enforcement activities. Areas of operation include the border environment, both land and sea, to observe and interdict unlawful crossings of persons and goods, the airspace surrounding defined DHS National Special Security Events or critical venues, and populated or unpopulated areas that are the subject of defined law enforcement activity or investigation. CBP's helicopter fleet operates out of 30 locations maintained by OAM across the United States.

*Fixed-Wing Aircraft*

CBP has manned fixed-wing P-3 AEW/LRT Orion aircraft operating out of specific operations centers in Corpus Christi, TX and Jacksonville, FL. CBP practices a defense in depth strategy of the borders of the United States and in active prosecution of attempts to smuggle persons or contraband by extending surveillance over international and coastal waters. As part of this strategy and as a means of integrating with the overall U.S. Government strategy to interdict the flow of narcotics and controlled substances across the U.S. southern borders, this defense in depth includes expanding the area of patrol to include the Caribbean and Eastern Pacific waters that border Source and Transit Zone countries.[1] Together the operations centers operate the P-3 aircraft primarily in Central and South America. Certain P-3s are used to intercept and track both aircraft and vessels for hours at a time while maintaining a covert standoff. CBP also operates several smaller, manned, fixed-wing aircraft out of OAM operational locations. These fixed-wing aircraft include piston-engine propeller-powered aircraft (Cessna models), larger turbo-prop powered aircraft (Bombardier Dash Eight, Pilatus, and Beechcraft Super King Air), and jet aircraft (Cessna Citation). These aircraft variously perform surveillance, tracking, interdiction, intercept, and information gathering roles. Fixed-Wing Aircraft employ various types of sensor technology including video, still, and radar images, and Law Enforcement Technical Collection (LETC) (electronic signals information across the electromagnetic spectrum).

*UAS*

A UAS encompasses an unmanned aircraft, digital network, and personnel on the ground who operate the aircraft. CBP currently owns and operates ten such aircraft. The UAS aircraft include the Predator B[2] and the maritime variant of the Predator B, the Guardian, which allows CBP to conduct missions in areas that are remote, too rugged for ground access, or otherwise considered too high-risk for manned aircraft or personnel on the ground. The aircraft are stationed and principally controlled at four locations: Sierra Vista, AZ (4 aircraft); Grand Forks, ND (2 aircraft); Corpus Christi, TX (2 aircraft); and Cape Canaveral, FL (2 aircraft). CBP's UAS operate in accordance within the Federal Aviation Administration (FAA) Certificate of Authorization (COA) process. CBP works with the FAA to develop the COAs to define airspace for UAS operation. Consistent with the primary mission for the UAS, these COAs, which are in effect for a period of two years, define airspace (altitude, latitude, and longitude (geography)) along the border and outside of urban areas to support CBP UAS flight operations. As the FAA develops its roadmap to integrate UAS into the National Airspace System (NAS)[3], CBP will adjust to these new requirements and continue to employ UAS in pursuit of its primary border security mission.

---

[1] Source and Transit Zone countries are those nations working in partnership with the United States to interdict the flow of narcotics and controlled substances to the United States through the Caribbean Basin and along the coastal waters of the eastern Pacific Ocean. http://www.whitehouse.gov/ondcp/transit-zone-operations.

[2] The General Atomics Aeronautical Systems MQ-9 Predator B is a mid-size Unmanned Aerial Vehicle (UAV) approximately thirty-six feet in length, with a maximum gross weight of 10,500 pounds and a wing span of sixty-six feet.

[3] See, *FAA Modernization and Reform Act of 2012*, Pub. L. No.112-95, sec. 331, 126 Stat. 11, 72, which mandates that the FAA prepare a roadmap to integrate UAS into the NAS by 2015.

## *Uses of Aircraft*

*Patrol*

CBP uses all of its aircraft to patrol different parts of the border based on the specific strengths of the different aircraft. CBP P-3s patrol in a 42-million square mile area of the Western Caribbean and Eastern Pacific, known as the Source and Transit Zone, in search of drugs that are in transit towards U.S. shores. The P-3's distinctive detection capabilities allow highly-trained crews to identify emerging threats well beyond U.S. land borders. By providing surveillance of known air, land, and maritime smuggling routes in an area that is twice the size of the continental U.S., the P-3s detect, monitor, and disrupt smuggling activities before they reach shore.[1] As part of this patrol responsibility, images and radar information obtained in detecting, monitoring, or supporting activities is collected and maintained either for direct case support or to permit historical trend analysis regarding smuggling routes.

Along both the northern and southern borders CBP also employs UAS and smaller manned aircraft to help agents detect, identify, apprehend, and remove individuals and contraband illegally entering the United States at and between Ports of Entry (POE). The COA defined airspace establishes operational corridors for UAS activity both along and within 100 miles of the border for the northern border, and along and within 25 to 60 miles of the border for the southern border, exclusive of urban areas. CBP helicopters and manned fixed-wing aircraft may operate in and around urban areas; however, the principal mission remains focused on those areas between the POE. Images, LETC, and radar information, specifically with respect to border areas between the POEs, are collected in support of case development or to permit trend analysis.

Following a flight, the images are provided to OIIL for processing, exploitation, and dissemination. Subsequently, and only upon request, OIIL provides access to the forensic analysis of a particular image and area to authorized persons who have a "need to know;" when the dissemination is in response to a particular law enforcement activity or case, that analysis may include PII.

Persons who are apprehended and who were video recorded from a UAS or a manned aircraft may have the video of their crossing and/or apprehension associated with a case file that contains their PII.

Separately, CBP also deploys manned fixed-wing aircraft with LETC sensors over the border area in support of its counter-terrorism and interdiction of smuggling operations. The LETC sensors permit surveillance of the electromagnetic spectrum for the purpose of identifying organized border crossing activity between the ports of entry.

*Investigative Operations*

CBP uses both UAS and manned aircraft in support of other DHS components, such as U.S. Immigration and Enforcement (ICE), or other federal law enforcement agencies, such as

---

[1] The Anti-Drug Abuse Act of 1988 established the Office of National Drug Control Policy (ONDCP) to set priorities, implement a national strategy, and certify Federal drug-control budgets. Interdiction of the flow of illicit drugs through the Source and Transit Zone is a critical component of the National Drug Control Strategy prepared annually by ONDCP.

the Federal Bureau of Investigation (FBI) or Drug Enforcement Agency (DEA). Requests for aircraft support that are related to the border surveillance must be directed to the Assistant Commissioner, OIIL, for authorization. Each request for information follows a standard process and is reviewed and considered in terms of the requesting agencies' authorities to receive the sought after information, CBP's own authority to lend assistance, and CBP's ability to integrate the information collection into its mission. Separately, OAM must determine the availability of aircraft type and the integration of the requested activity into its flight operations.

Typical support missions include overhead observation of previously identified persons, specified locations, and particular conveyances for enhanced situational awareness and increased officer safety. For example, the UAS could conduct surveillance over a building to inform ground units of the general external layout of the building or provide the location of vehicles or individuals outside the building. When flying a UAS in support of another component or government agency for an investigative operation, CBP may provide the other agency with a direct video feed through access controls or with a downloaded video recording of the operation, in whole or in part, based on the request. Similarly, CBP may deploy a helicopter or manned fixed-wing aircraft to provide over top visibility into a developing incident. Video images from the Electrical Optical/Infrared ball (EO/IR) ball are fed through the DHS firewall to "Big Pipe," a video and image distribution network operating within the CBP/DHS firewall, to identified users, analysts, and decision makers for real-time mission support and border protection.

*Disasters*

The P-3 may be used to conduct reconnaissance missions during natural disasters in support of FEMA. During these missions, P-3s can provide near real-time, high quality video of affected areas to first responders and FEMA. P-3s are equipped with similarly capable EO/IR Ball cameras; the images are also fed through a transmission to a ground station where the video is decrypted and fed to Big Pipe to disseminate inside the DHS firewall to authorized users within DHS and any other requesting agency.

UAS may also be used outside existing COAs during natural disasters once the government has issued a disaster declaration. For example, the UAS may fly missions in support of other government agencies such as the National Oceanic and Atmospheric Administration (NOAA) or FEMA to provide video or radar images of flooding. In disaster situations, CBP works with the FAA to construct a COA defining the airspace where a CBP UAS may operate. The UAS may provide a real-time feed during flight through Big Pipe or, subsequently, an analyzed image comparing the raw feed to an image with identified details, noting changes, to FEMA, state emergency operations centers, United States Geological Survey (USGS), and/or the Army Corps of Engineers. Video from these operations are not used to identify individuals. As with other requests for support, disaster area overflight requests are assigned in accordance with the national policy regarding the tasking of CBP air assets.

*Officer Safety and Support to State and Local Law Enforcement*

State and local law enforcement officials may request aircraft support (e.g., UH-60, P-3, UAS) in emergency situations; often this involves circumstances when officer safety is implicated, and in which aerial surveillance is necessary or the terrain would be too difficult for law enforcement personnel to navigate. OIIL reviews each request to determine whether to respond and OAM reviews how and in what context it may respond. Based on both organizations within CBP, a decision is made whether to provide assistance. Access to video taken during emergency situations may be provided, either at a DHS/CBP facility or by temporarily granting direct access through the DHS firewall. Sharing of this information with state, local, or other government agencies is on a case by case basis as determined through CBP's Request for Information process.

As in the mission uses discussed above, UAS and manned aircraft offer several options for deploying information gathering equipment. The UAS can serve as force multiplier insofar as the UAS enables the monitoring of large areas of land more efficiently and with fewer personnel than other aviation assets. UAS can enhance situational awareness and increase officer safety by providing aerial support to officers on the ground by monitoring a fixed location while flying at a high altitude to reduce the likelihood of detection. Manned aircraft offer the ability to fly in more congested airspace and to transport officers, agents, equipment, and seized assets.

## Technology on Board the Aircraft

The various aircraft have different types of surveillance technology. Most aircraft, manned and unmanned have an EO/IR ball attached to provide a means of collecting information. The EO/IR ball installed on the UAS also assists the pilot during take-off and landing. While the cameras on each aircraft are not identical, they have almost identical performance specifications. The EO/IR ball is a camera, which employs a fixed-focus lens, that is capable of providing video at any altitude and allows operators, using digital zooming (software based image enhancement), to take small-scale aerial video images of buildings, vehicles, and people. Aircraft altitude directly affects a fixed-focus camera's performance; the higher the aircraft's altitude, the less detail an operator is able to see.

A lower altitude permits the EO/IR ball to provide greater detail in an image, which may permit identification; this observation activity, however, does not occur unnoticed or subject to attempts at evasion, and therefore is more often part of a defined law enforcement operation. Persons are often successful at hiding their identity from known surveillance aircraft by simply looking away.

At present, the flight and mission parameters for the UAS place their operation within an altitude block of 19,000 to 28,000 feet, thereby effectively limiting the altitude for the EO/IR ball on a UAS to a minimum of 19,000 feet. At this minimum altitude, the camera does not provide enough detail for an operator to identify a person (that is to discern physical characteristics such as height, weight, eye color, hair style, or a facial image). The camera operator may have enough detail to identify whether an individual is carrying a long gun or wearing a back pack. At an altitude of 19,000 feet the camera operator cannot read a license plate, nor are license plate readers effective.

Conversely, the flight parameters for helicopters and fixed-wing aircraft are broader in terms of altitude and geography; their flight operations are integrated into the NAS and do not require a COA. The mission parameters and physical capabilities for helicopters and manned fixed-wing aircraft, however, place different operational restrictions upon the aircraft.

The EO/IR ball can provide daytime or nighttime visual video observation of movement or objects on the ground. The images, depending upon the aircraft deploying the camera, tend to be small in scale, to provide environmental context. A principal purpose for tracking a person or vehicle from an aircraft with an EO/IR ball is to assist CBP or law enforcement personnel on the ground with information to permit a safe encounter—this requires environmental context more than a best possible close-up of a face. When viewing vehicles, an operator can distinguish a car from a truck, and depending on the altitude at which the aircraft is flying, may be able to identify the model of the vehicle. During daytime flights, an operator may also be able to determine the color of the vehicle. The images of vehicles and/or individuals recorded by the EO/IR ball are not associated with any biographical information unless the individual is apprehended, at which point the video may be associated with the Personally Identifiable Information (PII) contained within the individual's case file.

In addition to EO/IR CBP deploys a UAS stationed along the Southwestern border in Sierra Vista, AZ, with the Wide Area Surveillance System (WASS). WASS uses a sensor mounted to the wing of a UAS to sweep large areas of border territory (approximately six kilometers in width) as the aircraft moves along its flight path. WASS alerts CBP to the existence of persons and/or vehicles along the border and provides coordinates to determine their location. The UAS pilot and sensor operator can then inform ground units of the location so that Border Patrol may coordinate an interdiction of the persons or vehicles. WASS provides a radar sensor image, which CBP may share through Big Pipe during operation.

Some manned and unmanned aircraft are also equipped with synthetic aperture radar that can provide black and white images in all weather. This radar can provide silhouettes of people and vehicles, but provides no identifying details. Using this technology, an operator is not able to pick up identifying characteristics of a person or a vehicle. The synthetic aperture radar is primarily used for change detection. For example, the operator can identify tire tracks on the ground that were not present in prior images provided by the radar. Similarly, an operator can use the synthetic aperture radar to determine the extent of flooding in a particular region by noting the changes to the topography.

Certain manned fixed-wing aircraft deploy LETC sensors used to detect electronic signals in the electromagnetic spectrum. These specifically designed aircraft operate in support of counter-terrorism efforts and to interdict organized smuggling (people, contraband, and controlled substances) operations within the border area. Like with the EO/IR ball, information from LETC sensors may be employed to support officers and agents on the ground as they move to a position where they can safely encounter observed persons. LETC aircraft sensors are solely deployed on manned fixed-wing aircraft.

Data on the digital video recorders on CBP aircraft are maintained for a maximum of 30 days and then overwritten by new data. The images and related data from CBP aircraft, both manned and unmanned, are provided through Big Pipe to identified users, analysts, and decision makers for real-time mission support and border protection. Images from the EO/IR ball mounted on the UAS are sent by an encrypted transmission, first to the satellite providing the control signals, and then, again by encrypted transmission, to the ground control station where the pilot and sensor operator are located. The image data is decrypted and brought inside the DHS firewall at the ground control station, where Big Pipe can ingest the data and provide a feed to assigned users and analysts.

Big Pipe is a fully distributed network hosted by CBP and supports not only event-based law enforcement missions, but also FEMA's National Response Framework. Big Pipe employs role-based access controls to provide users possessing a need to know access to distinct video feeds at command centers, other CBP/DHS locations, and for authorized persons with technical access through the DHS firewall. OAM retains control over defining users for Big Pipe and assigning access. After the creation of live mission data, Big Pipe manages the transmission, processing, distribution, consumption, and storage of the live mission data. Big Pipe archives selective mission data on a Big Pipe server hard drive for a maximum of 7 days, after which the data is deleted. Big Pipe does not use PII to retrieve stored mission data. Stored data is retrieved based on the date and time of the mission and only by authorized users on a need to know basis. If data is used for investigative purposes, and associated with a particular individual it goes into a case management system, which is covered by the corresponding Privacy Act System of Records Notice (SORN) for the case management system. Big Pipe, separately, provides a feed of video and radar images from UAS to the Air and Marine Operations Center (AMOC), where OIIL operates one of several PED cells to review this data over time to perform trend analysis and change detection. Video and radar images maintained by a PED cell, such as at the AMOC, are stored on a separate server dedicated to the PED cell mission for up to five years. The analyzed images may be shared by OIIL in response to law enforcement needs.

## Summary of Privacy Risks

The use of these aircraft and accompanying surveillance technologies presents several privacy concerns. The first concern is ensuring that CBP's collection and use of data from aerial surveillance remains within the scope of its authorities to protect the border and provide support for law enforcement activities, while continuing to preserve a person's right to privacy. CBP's border security mission has a broad mandate to determine the admissibility of persons and ensure that goods are not introduced into the United States contrary to law. Similarly, the statutory language in CBP's annual appropriations directs CBP Air and Marine to provide integrated and coordinated border interdiction and law enforcement support for homeland security missions, including assistance to federal, state, and local agencies and emergency humanitarian efforts; to provide airspace security for high-risk areas or National Special Security Events ; and to combat efforts to smuggle narcotics and other contraband into the United States . Deploying OAM's various air assets to support these missions improves DHS/CBP's capability to obtain streaming video, and to assess critical infrastructure before and after events.

CBP's use of manned and unmanned aircraft to conduct aerial observations is consistent with CBP's authorities and obligations. To the extent that aircraft flying in support of tactical operations overfly private residences, there is a minimal risk that a person's privacy might be unintentionally violated. The images captured are not personally identifiable without further investigative information. Neither manned nor unmanned aircraft physically intrude upon or disturb the use of private property. Further, the cameras deployed on UAS or manned aircraft do not have the capability to see through walls or otherwise collect information regarding what occurs in the interior of a building, nor is that their purpose. UAS operate primarily at an altitude between 19,000 and 28,000 feet pursuant to their COA approved by the FAA, and are focused as previously described.

A second privacy concern, specific to UAS, is that they present a perceived risk to privacy because they are able to fly for longer hours than manned aircraft and conduct surveillance undetected. Like other aircraft, UAS are useful for monitoring remote land border areas where patrols cannot easily travel and infrastructure is difficult or impossible to build. Unlike manned aircraft, UAS are operated by personnel on the ground, allowing the crew to be relieved while the UAS is still in the air. This capability allows UAS to provide long-range surveillance for greater lengths of time than manned aircraft. Because of their small size compared to manned aircraft, and the altitude at which UAS can operate, these physical attributes may serve to conceal the presence of a UAS and reduce detection of their operating noise while still being able to maneuver over a small area and provide surveillance. Other OAM operated long range fixed-wing aircraft cannot steadily monitor a set location because of their size and turning radius. Helicopters are more easily detected because of their noise and lower operational altitudes. This means that, unlike fixed-wing aircraft and helicopters, UAS can monitor either a moving target or a fixed location for relatively longer periods of time without the likelihood of detection.

While UAS can fly for longer periods of time, they are equipped with the same technology to conduct surveillance that is presently deployed on CBP manned aircraft. The only sensor available on UAS that is not used by CBP manned aircraft currently is the WASS sensor. The WASS sensor can only detect the presence of a person and track his or her movements (much the same way other radar technology can detect an object and track its movement); it cannot be used to identify a person. The WASS sensor is designed to sweep large areas of land and is only used to patrol along the southwest border and to assist with interdictions. Other technologies on the UAS are shared by CBP's manned aircraft. Putting these technologies on a UAS only enhances CBP's ability to perform its existing functions. For instance, CBP's surveillance video of a location used to smuggle persons or contraband using a UAS instead of a P-3 may be longer in duration with less interruption and less likelihood of detection.

To mitigate the risk presented by longer sustained surveillance of an individual or residence without the individual's knowledge, CBP has strict mission priorities for UAS and all aircraft operations. For instance, CBP aircraft may only be used in support of an authorized mission or investigation, the video or other data collected from CBP aircraft may only be accessed by authorized personnel with an authorized need to know, and the CBP-held video or other data is controlled through chains of custody and stored in secure locations until it is destroyed. In addition, the FAA requires CBP to construct a COA, in the instance of deploying a UAS, for a duration determined by the investigative activity or emergency circumstance, before conducting an operation away from the border and already established COAs.

The third privacy concern, unique to UAS, pertains to the security of the system itself and the potential for hijacking of the unmanned aircraft. CBP has taken several steps to protect UAS against potential hackers. All UAS are controlled and monitored at all times by operators in ground control stations using satellite communication that is relayed through an encrypted data feed. The ability to interfere with such an encrypted data feed requires disrupting the signal from satellite to UAS, for the purpose of acquiring the data feed or controlling the UAS. In the event that the ground control station loses its ability to control the UAS, another ground control station can pick up control of that UAS. The UAS use redundant navigation systems and GPS receivers so that if a signal is lost or someone attempts to override the signal, the UAS relies on these other systems and the GPS receivers for flight operations. In order to protect the airspace, the FAA is notified immediately if a UAS loses its signal. Furthermore, if communication between ground control and the UAS is ever interrupted or lost, the UAS are pre-programmed to fly to a precoordinated point in a remote location to orbit while waiting for the signal to be reestablished, or to continue to orbit this Flight Termination Point until the aircraft runs out of fuel and crashes.

Because of the unique privacy concerns raised by CBP's use of Aircraft Systems, CBP has conducted this PIA to evaluate the privacy risks associated with the use of Aircraft Systems and to enhance public understanding of the authorities, policies, procedures, and privacy controls related to that use.

# Fair Information Practice Principles (FIPPs)

The Privacy Act of 1974 articulates concepts of how the Federal government should treat individuals and their information and imposes duties upon Federal agencies regarding the collection, use, dissemination, and maintenance of personally identifiable information. Section 222(2) of the Homeland Security Act of 2002 states that the Chief Privacy Officer shall assure that information is handled in full compliance with the fair information practices as set out in the Privacy Act of 1974.

In response to this obligation, the DHS Privacy Office developed a set of Fair Information Practice Principles (FIPPs) from the underlying concepts of the Privacy Act to encompass the full breadth and diversity of the information and interactions of DHS. The FIPPs account for the nature and purpose of the information being collected in relation to DHS's mission to preserve, protect, and secure.

DHS conducts Privacy Impact Assessments on both programs and information technology systems, pursuant to the E-Government Act of 2002, Section 208, and the Homeland Security Act of 2002, Section 222. Given that Aircraft Systems and their associated devices are mechanical and operational systems rather than a distinct information technology system or collection of records pertaining to an individual that would be subject to the parameters of the Privacy Act, this PIA is conducted to relate the use of these observation and data collection platforms to the DHS construct of the FIPPs. This PIA examines the privacy impact of Aircraft Systems operations as it relates to the DHS FIPPs.

## 1. Principle of Transparency

Principle: *DHS should be transparent and provide notice to the individual regarding its collection, use, dissemination, and maintenance of PII. Technologies or systems using PII must be described in a SORN and PIA, as appropriate. There should be no system the existence of which is a secret.*

This PIA provides a level of transparency to the public about the current surveillance programs undertaken by CBP. The video, still images, signals information, and/or radar images do not clearly identify individuals. The only information about individuals that is collected and/or retained is the indication of a human form. These images, however, may be associated with a person if the person is apprehended. For example, video collected by an EO/IR ball may show several individuals traversing the land border and being intercepted by officers or agents of CBP. While the video resolution or radar mapping images are not sufficiently precise to permit actual identification, the circumstances of CBP interdiction and apprehension of a suspect in conjunction with the aerial surveillance are sufficient to link the indistinct images of persons traversing the ground to the case file. Individuals who are apprehended by CBP as a result of observation by aircraft at or near the border may have video of their crossing and apprehension associated with their enforcement case file. CBP obtains biographical data pertaining to the apprehended person at the moment of apprehension. CBP stores all biographical information obtained from apprehended individuals and any video or radar images of their movement obtained from the aircraft in the appropriate law enforcement case management system.

When CBP associates video, still images, signals information, and/or radar images with an individual after apprehension, that information becomes subject to the requirements of the Privacy Act in the same manner and to the same extent that the apprehension of the individual becomes a record in a Privacy Act system. The Privacy Act requires that agencies publish a SORN in the Federal Register describing the nature, purpose, maintenance, use, and sharing of the information. This PIA serves as notice to the public that information captured by Aircraft Systems may become subject to the Privacy Act once it is associated with an individual. Additionally, the video images associated with an individual's case file are covered by the appropriate law enforcement case management SORN, which maintains the case file. CBP will periodically re-assess the means by which the images from the aircraft are retrieved to determine whether the requirement for a SORN is triggered.

## 2. Principle of Individual Participation

Principle: *DHS should involve the individual in the process of using PII. DHS should, to the extent practical, seek individual consent for the collection, use, dissemination, and maintenance of PII and should provide mechanisms for appropriate access, correction, and redress regarding DHS's use of PII.*

Individual participation provides complementary benefits for the public and the government. The government is able to maintain the most accurate information about the public, and the public is given greater access to the amount and uses of the information maintained by the government. A traditional approach to individual participation is not always practical or possible for CBP, which has law enforcement and national security missions. Aircraft are primarily used to sweep the border area to locate individuals who are crossing the border illegally. Allowing an individual to consent to the collection, use, dissemination, and maintenance of video, still images, and/or radar images would compromise operations and would interfere with the U.S. government's ability to protect its borders, thereby lessening overall homeland security.

Individuals do not have the opportunity to restrict CBP's ability to collect information in the public sphere. Any information associated with an individual is part of a case file that is created as part of a law enforcement investigation or encounter. Providing individuals of interest access to information about them in the context of a pending law enforcement investigation may alert them to or otherwise compromise the investigation. Consequently, there is no mechanism for correction or redress for the video collected by the aircraft. Once that video is associated with an individual's case file, the individual must follow the procedure outlined in the corresponding privacy documents for that system. While individuals cannot participate in the initial collection of this information, they may contest or seek redress through any resulting proceedings brought against them. More information on redress is provided below.

## 3. Principle of Purpose Specification

Principle: *DHS should specifically articulate the authority which permits the collection of PII and specifically articulate the purpose or purposes for which the PII is intended to be used.*

The purpose specification principle requires DHS to 1) articulate the authority to retain the PII in question; and 2) articulate the purpose(s) for which DHS uses the PII.

CBP is authorized to collect video, other images, signals information, and data using aircraft in support of its border security mission and pursuant to the appropriations language mandating support for law enforcement as part of the mission of CBP Air and Marine. Together, these authorities allow CBP to obtain information in support of border interdiction of narcotics and other contraband, the prevention of the illegal entry of aliens into the United States, the security of airspace for high-risk areas or National Special Security Events, and in support of federal, state, and local law enforcement, counterterrorism, and emergency humanitarian efforts.

CBP may use video, still images, signals information, and/or radar images, obtained from aircraft, to apprehend individuals and to provide evidence of an illegal border crossing or other violation of law. Consistent with applicable laws and SORNs, the information may be shared with other state, local, federal, tribal, and foreign law enforcement agencies in furtherance of enforcement of their laws.

Video, still images, and/or radar images collected during investigative operations as part of a law enforcement investigation are used for enhanced situational awareness and increased officer safety, and may be used to provide evidence of a violation of law. These images are maintained in association with the investigative or case file that they support; their retention is managed by the same SORN and follows the handling of the investigative or case file.

Video, still images, and/or images collected in natural disaster and/or emergency situations are used for relief work and disaster reconnaissance. CBP typically provides a direct feed of the video captured by aircraft in these scenarios to provide support to FEMA or state emergency operating centers. Video, still images, and/or radar images are not associated with an individual and are only used to indicate where an individual or group of individuals may be for emergency response purposes.

## 4. Principle of Data Minimization

Principle: *DHS should only collect PII that is directly relevant and necessary to accomplish the specified purpose(s) and only retain PII for as long as is necessary to fulfill the specified purpose(s). PII should be disposed of in accordance with DHS records disposition schedules as approved by the National Archives and Records Administration (NARA).*

CBP seeks to minimize the collection and retention of video, signals information, and radar to that which is necessary and relevant to carry out CBP's mission. Accordingly, when aircraft are flown to patrol the border, they are authorized to fly the designated border surveillance mission area to ensure they are only capturing images and information necessary to detect, identify, apprehend, and remove persons and their possessions illegally entering the United States at and between POE. When aircraft are flown for investigative operations, officer safety incidents, or natural disaster reconnaissance, CBP approves and defines the specific mission that is authorized, and in the case of UAS, works with the FAA to construct a COA to establish airspace for that specific UAS operation. The video (that has not been associated with a case) remains on the digital video recorder originally used for recording until it over-written through re-use, which is after approximately 30 days.

After the creation of live mission data, Big Pipe manages the transmission, processing, distribution, consumption, and storage of the live mission data. Big Pipe archives selective mission data on a Big Pipe server hard drive for a maximum of 7 days, after which the data is deleted. Big Pipe does not use PII to retrieve stored mission data.

The information collected by the aircraft is not subject to the Privacy Act unless it is retrieved by using an individual's name or other unique identifier. If an individual is apprehended by CBP as a result of observation by aircraft or subsequent association from the presence of CBP assets, CBP may have video of that individual's apprehension associated with his or her enforcement case file. That video is retained according to the retention schedule of the SORN of the corresponding case management system. Video and Radar images obtained from UAS patrols of the border are also provided to PED cells operated by OIIL for use in analyses and intelligence products concerning historical, change detection (e.g., natural and man-made alterations to geography) along the border, and patterns of movement of persons across the border. This unassociated data, in conjunction with meta-data (such as latitude, longitude, date, and time of the imagery) is retained for a maximum of five years.

### 5. Principle of Use Limitation

Principle: *DHS should use PII solely for the purpose(s) specified in the notice. Sharing PII outside the Department should be for a purpose compatible with the purpose for which the PII was collected.*

CBP only collects video and/or radar images, and signals information via aircraft pursuant to its law enforcement authority, as part of its border security mission, or when flying a mission in support of another agency, and when that other agency's authority covers the mission either through delegation of authority or direct control of the information collected. For example, CBP has provided support to the U.S. Forest Service in response to large scale wild fires to permit an overview of the extent and scale of the fire and identification of hot spots; this activity is pursuant to a request from the Forest Service, is performed pursuant to their authority, and the images are conveyed through designated access to the Big Pipe video distribution service. While the video resolution, radar mapping images, and signals information are not sufficiently precise to permit actual identification of a person, the images or information may be associated with an individual from context within the image, circumstances surrounding the activity occurring in the image, or additional information obtained directly from the person by an officer or agent. The images or information are only associated with an individual if the individual is apprehended or if the images are taken as part of an ongoing law enforcement investigation. Accordingly the data can only be used for the purposes specified in section 3 of this PIA.

CBP has procedures and processes in place for sharing any data collected by aircraft, including when that information becomes associated with a case and is used as evidence against an apprehended individual. In addition, all requests for aerial surveillance for intelligence gathering purposes must receive prior approval by the Assistant Commissioner,

OIIL, before the air asset can conduct the flight. Similarly, requests for analytical products incorporating historical analysis of the border topography must be approved by the Assistant Commissioner, OIIL.

### 6. Principle of Data Quality and Integrity

Principle: *DHS should, to the extent practical, ensure that PII is accurate, relevant, timely, and complete, within the context of each use of the PII.*

As explained in section 4 (above), to ensure that the PII captured by aircraft is relevant and timely, any video, still images, signals information, and/or radar images must be associated within 30 days with the individual CBP apprehends, or the video/digital image is overwritten by OAM. Video and/or radar images are of no continuing value in a law enforcement support context unless they are associated with an individual during an apprehension because the video resolution or radar mapping images are not sufficiently precise to permit actual identification of individuals. Video and/or radar images that are not associated with a person provide value in an intelligence context for helping to demonstrate the state of change occurring over time along the border. These unassociated images are separately maintained by OIIL for a maximum of five years.

To preserve the quality and integrity of the information collected that is used as evidence, CBP requires its officer/agents to successfully complete training on the proper operation of the recording equipment on its aircraft. The training includes correct techniques to copy recorded evidence from a non-portable hard drive to portable digital media and procedures to ensure that such evidence is not co-mingled with data from other investigations. The training also includes procedures to maintain an adequate chain of custody for all recorded evidence. Each officer/agent making a recording must ensure that the time and date shown in the original recording is accurate. After a mission is completed, the officer/agent must ensure that the original record is transferred entirely, in its original format, to portable media. The transferred data must not be edited or altered in any way. The officer/agent making the recording must label all copies of portable media with the corresponding case number (if available), the date and place of the original recording, and the names of the officer/agent and aircraft commander. The officer/agent making the recording must also label, initial, and maintain possession of the evidence until custody is properly transferred to the appropriate designated evidence custodian, case agent, Assistant United States Attorney, or other appropriate government official. As with any information associated with a case file, once the images are cross referenced to an investigation or case, they become covered by the system of records for that case file system and subject to the access and amendment provision of that system.

### 7. Principle of Security

Principle: *DHS should protect PII (in all forms) through appropriate security safeguards against risks such as loss, unauthorized access or use, destruction, modification, or unintended or inappropriate disclosure.*

CBP has taken steps to protect live video feeds, signals information, and recorded video, radar, and/or still pictures captured by its aircraft. Live video and flight information, which are sent from the UAS, are passed along an encrypted feed from the UAS through the satellite relay to the ground control station. Similarly, control information from the ground control station to the UAS also passes along an encrypted feed. Video and data transmitted in real time via Big Pipe, a closed system with restricted access, is subject to access controls and an approval process requiring clearance by one of two CBP/OAM system administrators to ensure that only authorized users with a need to know have access to the video feeds. The real time video feeds are not recorded and archived. Any recorded images that are saved to be used as evidence or for intelligence gathering must be handled in accordance with CBP policy. Images that are used as evidence must be handled according to the procedures detailed in section 6 of this PIA. All recorded evidence must be kept in a locked container, segregated from other property and/or equipment. Video that is collected during an investigative operation that contains sensitive analytical surveillance, or reconnaissance related data may not be disclosed unless a request for disclosure has been submitted to the OIIL Collections Division Director. The request must include a copy of the information that is to be disclosed, must clearly specify the name of the intended recipient, how the information will be used, and the reasons justifying the disclosure. In the event that the information is disclosed, the OIIL Collections Division Director or his/her designee is required to redact law enforcement sensitive information, PII, and other sensitive related data unless the requestor has a need-to-know.

### 8. Principle of Accountability and Auditing

Principle: *DHS should be accountable for complying with these principles, providing training to all employees and contractors who use PII, and should audit the actual use of PII to demonstrate compliance with these principles and all applicable privacy protection requirements.*

All CBP employees are required to complete annual privacy awareness training, in addition to training on ethics and the CBP Code of Conduct. Access controls, both physical and technological, are in place to ensure only authorized access to the aircraft systems and the collected data/images.

Moreover, CBP requires its employees to successfully complete training on techniques to copy recorded evidence to portable digital media and requires them to follow procedures to ensure that such evidence is not co-mingled with data from other investigations. Employees must follow procedures to maintain an adequate chain of custody in the event that the information is used as evidence.

OIIL has a process in place for restricting the dissemination of video, still images, and radar images and keeps a log of the disclosures. Also, OIIL redacts law enforcement sensitive information, PII, and other sensitive related data unless the requestor has a valid need-to-know. Separately, CBP periodically reviews the logs or disclosure records to ensure compliance with established privacy policies, practices, and procedures for associated systems.

## Conclusion

CBP operates aircraft systems in support of its border protection and law enforcement support missions. These systems provide a variety of mobile platforms from which to obtain signals information, video, still, and radar images of persons and vehicles in the border area or that are the subject of an investigation or law enforcement activity. The collection of these images and signals information complies with the same internal procedures and practices required of any surveillance using any means by CBP officers and agents. The distinct capabilities of the different aircraft operated by OAM enhance CBP's ability to conduct certain missions pertaining to information collection, surveillance, or reconnaissance; however, the processes and procedures for authorizing and accounting for how, when, and where information is obtained remain consistent with CBP's traditional border security and law enforcement practices and policy. As technology improves, operating environments change, and policies adapt, this PIA will be updated and amended to refresh the analysis of these changes on the privacy of persons, who directly or indirectly come into contact with the information and data collection activities associated with CBP Air operations.

## Responsible Officials

Lothar Eckardt
Executive Director, National Air Security Operations
Office of Air & Marine
U.S. Customs and Border Protection
202-344-3950

Laurence Castelli
CBP Privacy Officer
Office of Privacy and Diversity
Office of the Commissioner
U.S. Customs and Border Protection
202-325-0280

## Approval Signature Page

Original signed and on file with the DHS Privacy Office.

Jonathan R. Cantor
Acting Chief Privacy Officer
Department of Homeland Security

# Technology Decision Tool

## *Striking a Balance — Tool Provides Cost/Benefit Analysis for Public Safety*

As budgets shrink and public safety needs grow, agencies must keep a watchful eye on the bottom line. Although officer and community safety is always a priority, efficiency of operations is a key consideration when making purchasing decisions. The Technology Decision Tool helps agencies make safe and sound acquisitions.

The tool was developed by the National Law Enforcement and Corrections Technology Center with input from technology experts from both large and small agencies who have first-hand experience in successfully evaluating and implementing technology projects. It guides agencies through a customized cost/benefit analysis exercise to help them make the best decisions for their officers and their communities. It directs decision-making based on needs, availability of technology, and lifecycle costs of products and training. Safety considerations meet budget realities and viable solutions are offered and evaluated.

To access the Technology Decision Tool, go to https://www.justnet.org/pdf/Technology-Decision-Tool.pdf. (Sample screens below.)

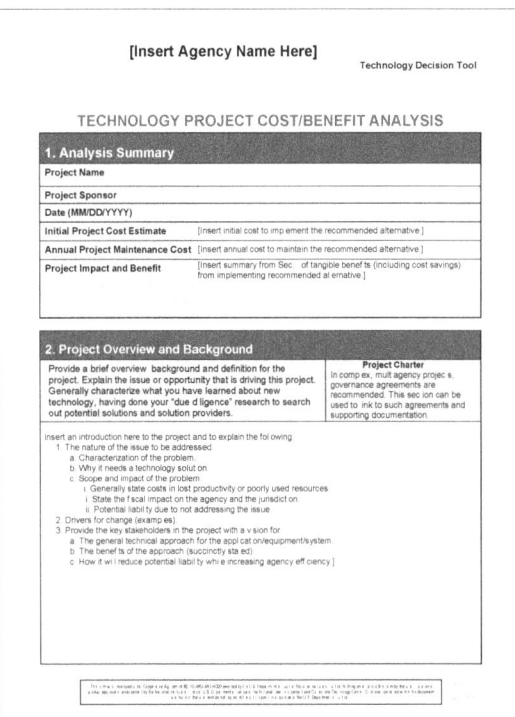

# Things to Consider for Unmanned Aircraft Systems (UAS) Procurement

With a heightened public concern about the use of UAS operations by government agencies, there are both programmatic and technical specifications that need to be addressed as part of the decision to purchase this technology. Even if you can justify the cost and use of a small UAS (sUAS), if your constituents are not convinced that this technology is less intrusive than current methods of aerial observation and/or surveillance, you could have a program with little public support.

**Why do we need an sUAS?**

- Remembering the mantra, "The mission must drive the technology as opposed to the technology driving the mission," can you clearly state the intended mission(s) that the UAS is slated to fill in your agency?

- Are there other ways in which your intended sUAS mission can be accomplished? Manned aircraft? Fixed site and/or pole cameras? Ground robot deployed cameras?

- Is there a positive return on investment (lifecycle costs of the UAS and payloads) that is measurable?

- Is the seller an established company that will be around to support the lifecycle of the UAS (5+ years)?

- Do you know the requirements to fly a UAS in the National Airspace (NAS) as established by the Federal Aviation Administration (FAA)?

- Does the seller know the requirements to fly a UAS in the National Airspace (NAS) as established by the FAA, which includes the requirements for conducting a sales demonstration flight?

- Have you conducted a search to see what UAS technologies are currently available for purchase and their respective costs?

- Are you able to clearly state to the public and the media the reason for the use of this technology and garner their support?

If, after answering the above questions, you still want to purchase a UAS, there are still a lot of technical specifications that need to be addressed:

**Specifications for Fixed-Wing UAS Procurement:**

- Weight and size of aircraft should be considered for costs, portability, current rules and regulations, and safety of operation.

- Battery powered (reliability and quiet).

- Gasoline powered (loud but longer flight time).

- Multiple batteries with quick change ability in the field.

- Easy and safe fuel (gasoline) in field.

- Ability to do quick battery charging in the field.

- Durable enough to be dropped from 10 feet without significant damage.

- Ability to quickly repair any damage in the field with included parts and tools.

- Delivered in a protected shipping case (hard plastic with foam innards) to be used for storage and deployment.

- Flight endurance of the aircraft should provide for mission goals and not hinder the ability to recover in a confined area. *(Note: 1) You should be aware that wind and humidity conditions create drag on the aerial vehicle, so an advertised flight time of 30 minutes may end up being 15 minutes of battery time for a mission; 2) The charge of the batteries decrease over time so if you start with a battery that holds a 30 minute charge, over the lifecycle of the battery your flight time will decrease naturally.)*

- Hand launchable.

- Box to flight time (setup time) in less than 20 minutes.

- Ability to preload flight plans with waypoint and altitude inputs.

- Lost-link that includes returned to land at a predefined location.

- Ability to track an image during flight to enable persistent monitoring of an object.

- Reliable GPS navigation.

- Automated stability during manual operation.

- Operating Manual.

- Maintenance Manual.

- Standardized operations training program for two pilots and two observers.

- Manufacturer provided itemized replacement part list.

- Multi-sensor capability (prefer EO/IR during the same flight).

- Live video streaming.

- Ability to record video in standard formats (MP3, MP4, JPEG, etc.).

- Extract video via memory card or thumb drive.

- Video includes date, time, and geographic location.

- Ability to safely fly in wind conditions that prevail in your normal operating area (ensure that the manufacturer can demonstrate safe operations in the winds that you expect in your normal operating areas).

- Will the aircraft be operated near or over water? If so, does the aircraft float? Is it waterproof and if so to what standard? Have the manufacturer demonstrate this feature.

- 5-year shelf-life minimum.

- Ability to upgrade the system and sensors.

**Specifications for Vertical Take-Off and Landing (VTOL) UAS Procurement:**

- Weight and size of aircraft should be considered for costs, portability, current rules and regulations, and safety of operation.

- Battery powered (reliability and quiet).

- Gasoline powered (loud but longer flight time).

- Multiple batteries with quick change ability in the field.

- Easy and safe fuel (gasoline) in the field.

- Ability to do quick battery charging in the field.

- Durable enough to be dropped from 10 feet without significant damage.

- Ability to quickly repair any damage in the field with included parts and tools.

- Delivered in protected shipping case (hard plastic with foam innards) to be used for storage and deployment.

- Flight endurance of the aircraft should provide for mission goals and not hinder the ability to recover in a confined area. *(Note: 1) You should be aware that wind and humidity conditions create drag on the aerial vehicle, so an advertised flight time of 30 minutes may end up being 15 minutes of battery time for a mission; 2) The charge of the batteries decrease over time so if you start with a battery that holds a 30 minute charge, over the lifecycle of the battery your flight time will decrease naturally.)*

- Box to flight time (setup time) in less than 20 minutes.

- Ability to preload flight plans with waypoint and altitude inputs.

- Lost-link that includes returned to land at a predefined location.

- Ability to track an image during flight to enable persistent monitoring of an object.

- Reliable GPS navigation.

- Automated stability during manual operation.

- Operating Manual.

- Maintenance Manual.

- Standardized operations training program for two pilots and two observers.

- Manufacturer provided itemized replacement part list.

- Multi-sensor capability (prefer EO/IR during the same flight).

- Live video streaming.

- Ability to record video in standard formats (MP3, MP4, JPEG, etc.).

- Extract video via memory card or thumb drive.

- Video includes date, time, and geographic location.

- Ability to safely fly in wind conditions that prevail in your normal operating area (ensure that the manufacturer can demonstrate safe operations in the winds that you expect in your normal operating areas).

- Will the aircraft be operated near or over water? If so, does the aircraft float? Is it waterproof and if so to what standard? Have the manufacturer demonstrate this feature.

- 5-year shelf-life minimum.

- Ability to upgrade the system and sensors.

**Spectrum:**

It is important to check that the radio frequencies that control flight of the UAS and/or controls payloads (cameras/sensors/downlinks/uplinks, etc.) are properly licensed and legal to use:

- The Federal Communications Commission (FCC) will authorize frequencies for control of UAS operations by non-federal entities (e.g., state and local government public safety agencies) consistent with the spectrum allocations and service rules for such uses.

- The Federal Communications Commission (FCC) will authorize frequencies for non-control operations on board UAS (e.g., video transmission, sensor data, etc.) by non-federal entities (e.g., state and local government public safety agencies) consistent with the spectrum allocations and service rules for such uses.

- Command and control (including video data) uplink; and video (air-to-ground) downlink frequencies for federal government UAS operations by the Department of Defense and other federal agencies (e.g., Interior, Homeland Security, etc.), will be authorized by the National Telecommunications and Information Administration (NTIA) consistent with radio frequency regulations and spectrum allocations for such uses.

The free Technology Decision Tool (https://www.justnet.org/pdf/Technology-Decision-Tool.pdf) can be used to assist in the decision to purchase an sUAS. If you have any questions, please contact Mike O'Shea, NIJ Senior Law Enforcement Program Manager, at michael.oshea@usdoj.gov.

## UAS Video

### *Eyes in the Sky: How Law Enforcement Uses Unmanned Aircraft Systems*

Some law enforcement agencies have employed the use of unmanned aircraft systems (UAS) to help investigate crime scenes, traffic accidents, missing person cases, and other emergencies.

This video, produced by the Justice Technology Information Center (JTIC), a program of the National Institute of Justice, Office of Justice Programs, U.S. Department of Justice, features the Mesa County, Colorado, Sheriff's Office and its six-year-old UAS program. It explores the equipment and procedures criminal justice professionals need to develop an effective and cost-efficient program of their own. Federal Aviation Administration regulations are discussed and facts are presented to address community concerns about safety and privacy.

To view the video, go to https://www.youtube.com/watch?v=2gzeXn_OoqI.

# NIJ Unmanned Aircraft Systems Expert Convening Agenda

*The National Institute of Justice's Unmanned Aircraft Systems*
*Expert Convening Agenda*

**Day One:**

**August 12, 2015**        Joint Session

Room 3101

810 7th Street, N.W.

Washington, DC 20531

**8:30–9:00 a.m.**        **Registration/Networking**

**9:00–9:30 a.m.**        **Welcoming Remarks**

**Beth McGarry**

*Principal Deputy Assistant Attorney General*

*Office of Justice Programs*

and

**Nancy Rodriguez**

*Director*

*National Institute of Justice*

and

**Erika Brown Lee**

*Chief Privacy and Civil Liberties Officer*

*United States Department of Justice*

**9:30–10:30 a.m.**        **Setting the Stage: An Overview of Public Safety Unmanned**

**Aircraft Systems Programs**

*Moderator:* **Michael K. O'Shea**

*Senior Law Enforcement Program Manager*

*National Institute of Justice's Office of Science and Technology*

9:30–9:45 a.m.

**Major General Randolph Alles (USMC Ret.)**

Assistant Commissioner for the Office of Air and Marine

U.S. Customs and Border Protection

U.S. Department of Homeland Security

Washington, DC

9:45–10:00 a.m.

**Captain Tom Madigan**

Alameda County Sheriff's Office

Office of Homeland Security and Emergency Services

Dublin, CA

10:00–10:15 a.m.

**Tim Herlocker**

Director, Emergency Operations Center

Fire Department of New York

New York, NY

10:15–10:30

**Daniel Schwarzbach**

Executive Director/CEO

Airborne Law Enforcement Association

Frederick, MD

10:30–10:45 a.m.     **Break/Networking**

10:45–11:45 a.m.     **Setting the Stage: An Overview of the Issues of Community**
**Acceptance of Unmanned Aircraft Systems Use by Public Safety**

*Moderator:* **Maureen McGough, Esq.**

*Policy Advisor*

*National Institute of Justice*

10:45–11:00 a.m.

**Jay Stanley**

Senior Policy Analyst

ACLU Speech, Privacy & Technology Project

American Civil Liberties Union

Washington, DC

**11:00–11:15 a.m.**

**Jennifer Lynch**

Senior Staff Attorney

Electronic Frontier Foundation

San Francisco, CA

**11:15–11:30 a.m.**

**Mario Mairena**

Senior Government Relations Manager

Association for Unmanned Vehicle Systems International

Arlington, VA

**11:30–11:45 a.m.**

**Lisa Ellman**

Partner & Co Chair of Global UAS Practice

Hogan Lovells US LLP

Washington, DC

**11:45 a.m.–1:15 p.m.**    **Lunch on your own in the area**

**\* NOTE: Afternoon Sessions are in OJP's Rooms 8101 & 8109 (8th Floor Telecommute Center)**

**1:15 p.m.–4:30 p.m.**    **Public Safety Group (8101)**

*Moderator:* **Tom Frazier**

**1:15–2:05 p.m.**

**Session One:** Does our Agency need this Technology? A facilitated dialogue on the Pros and Cons of Public Safety UAS operations.

**2:05–2:20 p.m.**

**Break/Networking**

**2:20–3:10 p.m.**

**Session Two:** A facilitated dialogue concerning 1) The Presidential Memorandum on UAS; and 2) The FAA's proposed small UAS (sUAS) rule and its implications.

**3:10–3:25 p.m.**

**Break/Networking**

**3:25–4:15 p.m.**

**Session Three:** A facilitated dialogue concerning public safety UAS operations and their implications for privacy policies.

**4:15–4:30 p.m.**

**Day One Concluding Remarks and Information for Day Two**

Day Two:

August 13, 2015, Joint Session Room 3101 ('Fishbowl')

| | |
|---|---|
| 8:30–9:00 a.m. | Registration/Networking |
| 9:00–9:30 a.m. | Recap of Day One (Group Moderators) |
| 9:30–10:30 a.m. | Understanding Privacy and Legal/Liability Concerns |
| 10:30–10:45 a.m. | Break/Networking |
| 10:45–11:45 a. m. | Guidance for Community Engagement: Collaboration and Transparency |
| 11:45 a.m.–1:15 p.m. | Lunch on your own in the area |
| 1:15–2:15 p.m. | Developing Recommendations and Procedures |
| 2:15–2:30 p.m. | Break/Networking |
| 2:30–3:30 p.m. | Overcoming Obstacles: Prioritizing Issues to Determine Solutions |
| 3:30–3:45 p.m. | Break/Networking |
| 3:45–4:30 p.m. | Concluding Remarks and Next Steps |

# UAS Convening PowerPoint Presentations

## *Tim Herlocker*

*Director, Emergency Operations Center*
*Fire Department of New York*

Canyons

Dense Neighborhoods

Tree Lined Streets & Parked Cars

Island City

Areas of Dense Brush

Active Harbor

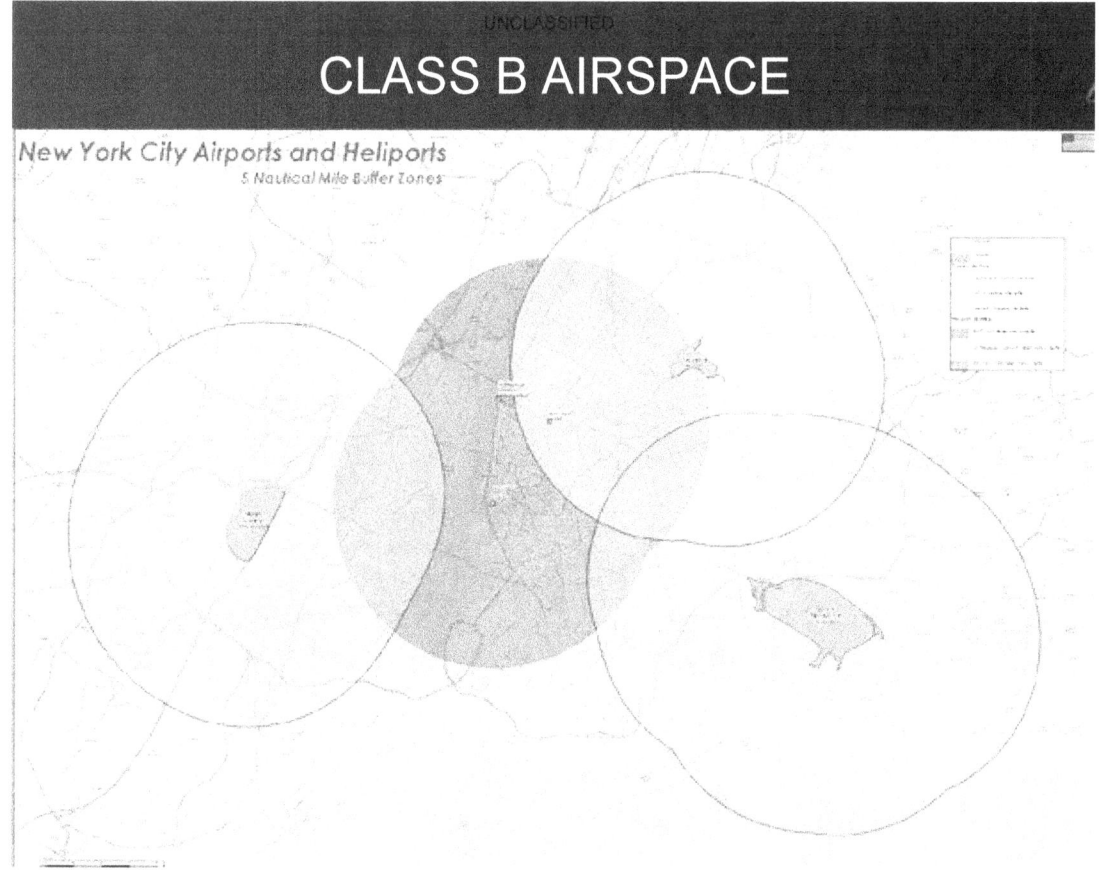

## Better Situational Awareness

- The 911 Lesson

- What's the roof look like?

- The rear of the building?

- Danger?

- Do other agencies need this?

UNCLASSIFIED

## The Oblique Aerial View

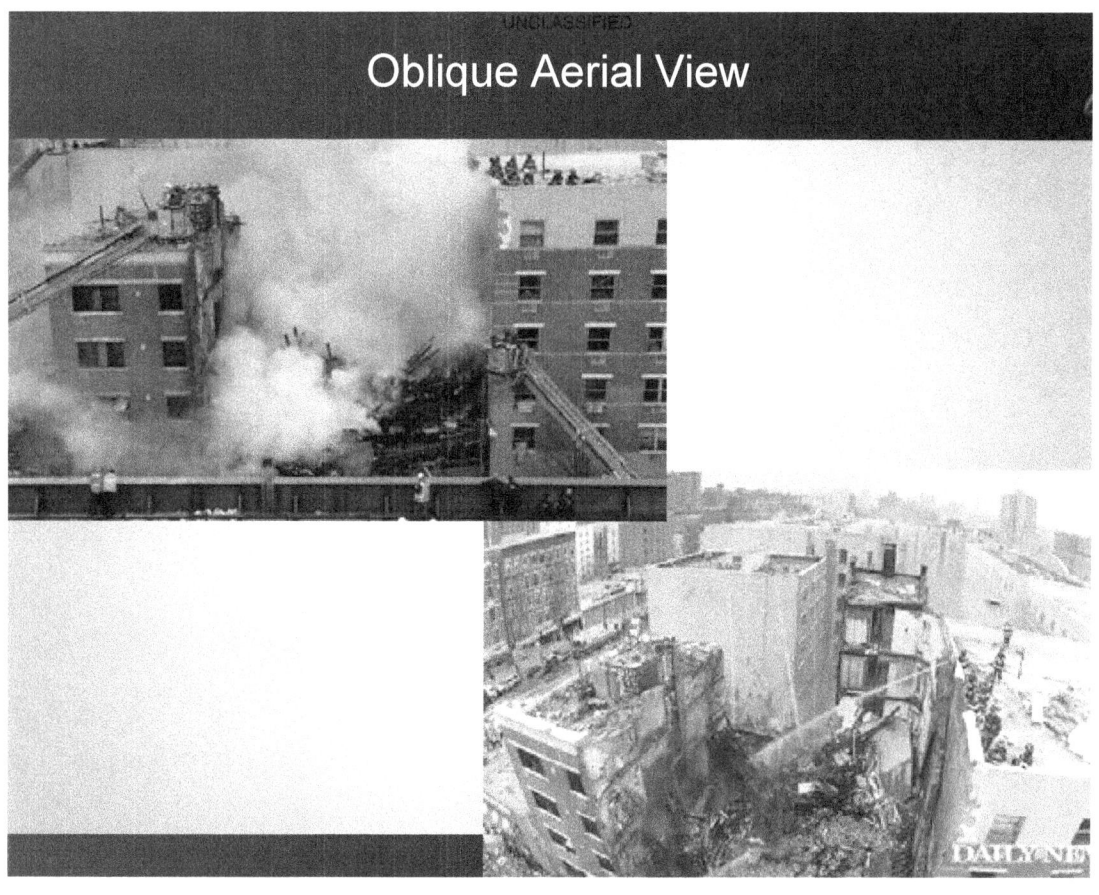

Oblique Aerial View

Oblique Aerial View

UNCLASSIFIED

# UAS CONOPS

- Operate at 2-Alarm and greater incidents occurring within Class B airspace on a 24/7 basis;

- AOR will include all five Boroughs and waterways extending beyond city limits;

- Anticipate UAS operation at 15 or more incidents per month, to include nighttime operations;

- Deployment point will be 1 to blocks from the incident location in order to capture an oblique view of the incident and not impede fire operations;

- FDNY will operate up to two UAVs at an incident, separated by a minimum of three blocks (600 to 1000 feet);

- FDNY will limit UAS operations to below 200 feet;

- UASs will be assigned to three separate units, located in different areas of the City in order to reduce response time to an incident;

- In order to maintain a 24/7 capability, the FDNY will require a minimum of 15 Department personnel, certified to operate a UAS;

- FDNY will certify its UAS "pilots" based on completion of FAA approved ground school or equivalent UAS training course.

UNCLASSIFIED

UNCLASSIFIED

# Tethering

UNCLASSIFIED

# UAS Requirements

- Maintain adequate tension/control of the tether to eliminate displacement of the tether;

- Receive power and control via the tether;

- Adequate tensile strength in the tether and anchor to prevent loss of the UAS;

- Maintain continuous flight for a minimum of six hours at 200 feet;

- Autonomously ascend, descend and maintain station within a tight radius of the anchor point for the tether;

- Station-keeping technology (other than GPS), and battery assisted return to base in the event of a loss of the tether;

- EO/IR camera

UNCLASSIFIED

UNCLASSIFIED

# Tethered UASs coming to market

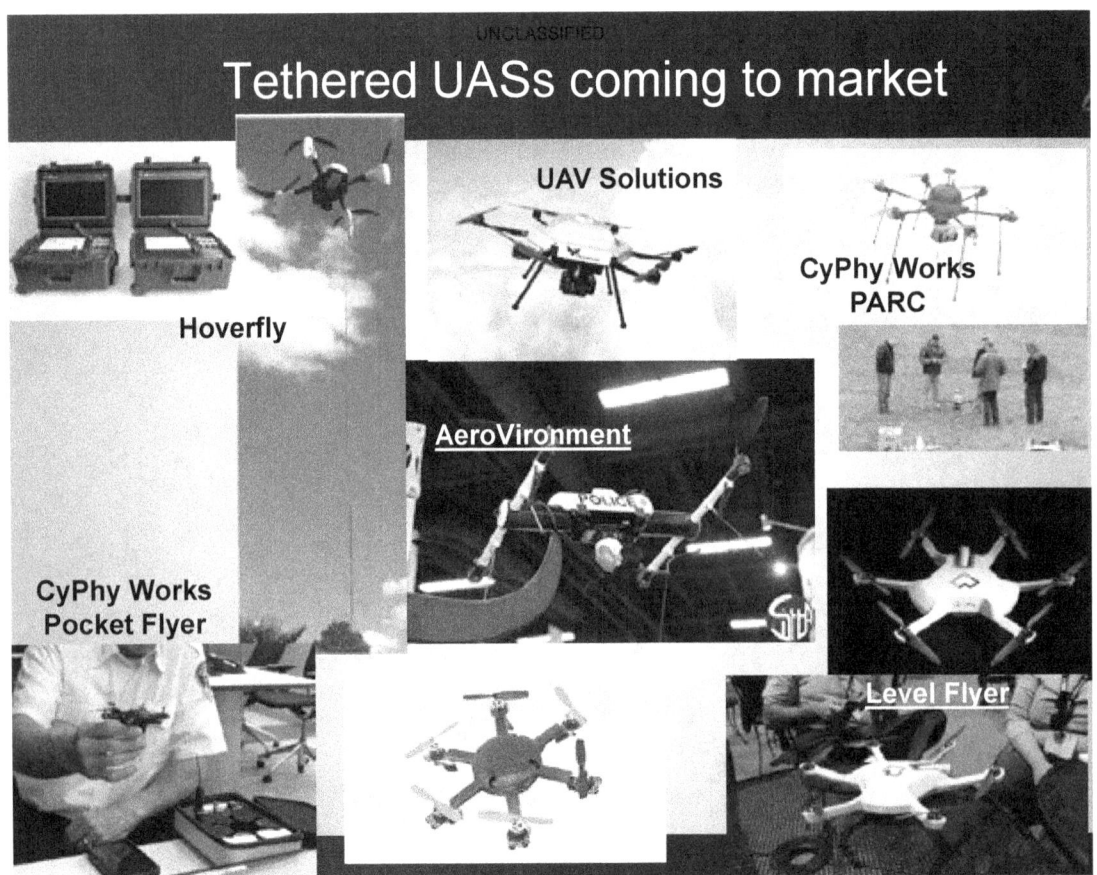

Hoverfly

UAV Solutions

CyPhy Works PARC

AeroVironment

CyPhy Works Pocket Flyer

Level Flyer

www.ingramcontent.com/pod-product-compliance
Lightning Source LLC
Chambersburg PA
CBHW081211180526
45170CB00006B/2300